GLOUCESTER DOCKS

An Illustrated History

Hugh Conway-Jones

ALAN SUTTON &
GLOUCESTERSHIRE COUNTY LIBRARY
1984

The County Library Series is published jointly by Alan Sutton
Publishing Limited and Gloucestershire County Library. All correspondence relating to the series should be addressed to:

Alan Sutton Publishing Limited
17a Brunswick Road
Gloucester GL1 1HG

First published 1984

BRITISH LIBRARY CATALOGUING IN PUBLICATION DATA

Conway-Jones,H.
   Gloucester Docks—an Illustrated History
   1.   Gloucester (Gloucestershire)—Docks,
   wharves, etc.—History
   I.   Title
   387.1'5   HE558.G5/

ISBN 0-86299-085-8

Typesetting and origination by
Alan Sutton Publishing Limited.
Photoset Melior 10/12.
Printed in Great Britain.

Contents

Maps

Acknowledgements

I welcome this opportunity of expressing my thanks to the many people who have helped me in my studies. I would particularly like to thank the staff of the Gloucester City Library and also the staff of the Gloucestershire Record Office. Nothing seems to be too much trouble for them, and they have brought to my attention many useful pieces of information. I also remember with gratitude the encouragement I received from Mr. Brian Smith, formerly at the Gloucestershire Record Office, whose evening class introduced me to the fascinating world of local history studies. My thanks are due to Mr. Anthony Done who has provided an invaluable index to pictures of Gloucester and who is now indexing the *Gloucester Journal* and also to the Reverend W. Awdry who made available his extensive notes on the Birmingham and Gloucester Railway. I have been very much encouraged in my studies by the interest and support shown by members of the Gloucester Civic Trust and the Gloucestershire Society for Industrial Archaeology. I also appreciate the assistance I have received from the staff of the British Waterways Board at Gloucester.

I have been very much helped in writing the later chapters by the recollections of many people who knew the docks when they were busy, with some memories going back to the earliest years of this century. I am particularly indebted to Mr. Alf Thomas with whom I have had many enjoyable conversations. I am also grateful to Mr. G.H. Allison, Mr. C. Boucher, Mr. H.G. Boucher, Mr. B.J. Cooke, Mr. W.H. Cullis, Mr. H. Done, Mr. T.R. Dury, Mr. W. Ellis, Mr. K. Gibbs, Mr. A. Gough, Mr. D.L. Harvey, Mr. G. Lippett, Mr. T.H. Mayo, Mr. G.C. Moody, Mr. R. Price, Mr. H.A. Roberts, Mr. K. Robins, Mr. W.G. Sargent, Mr. B.L.H. Shaw, Mr. A.G. Simmonds, Mr. S.W. Smith, Mr. W.J. Stokes, Mr. L. Stone, Mr. H.I. Williams, Mrs. C. Wixey and Mrs. R. Yates. I am grateful to Mrs. Sue Balcombe for doing the typing and to Mr. Neville Crawford for reading the typescript and making many helpful suggestions. Finally, I thank my wife and family for their tolerance and their support.

Hugh Conway-Jones,
Barnwood, Gloucester

Illustrations

The author is grateful to the following

Russell Adams F.R.P.S. Gloucester   159
Mr. H. Allison   131a
B.B.C. Hulton Picture Library   107b
Mr. D.E. Bick   19
British Waterways Board (Gloucester)   31 37a 81 97b 113 129a 146 150
Cheltenham Postcard Centre   103a
Mr. L.E. Copeland   63 135a 135b 137b 141a 143b
Mr. A.H. Done   122 133a 136 147b
Mr. W. Ellis   75
Mr. H.E. Etheridge   143a
Mr. B. Frith   88 89a 89b
Mr. K. Gibbs   147a 149b 152a 152b
Gloster Photographic Services   105b
Gloucester Citizen and Journal   151a 153 161
Gloucester City Library   12 15 25 33 39 41 51 55 57 65 74 85 86 87 92 95a 96 97a 99b 105a 115a 120 121b 123 125b 131b
Gloucester City Museum and Art Gallery   47
Gloucester Civic Trust and Gloucester City Museum and Art Gallery 11
Gloucester Chronicle   69
Gloucestershire Record Office   76 77 104 130 148
Mr. C.F. Godwin   103b
Mr. K.E.J. Hill   78
Martin Latham   27 35
Mr. E.A. Leah   127a
Mr. F.H. Lloyd   142
Mrs. F. Minett   95b
Mr. R.V. Morris   101 125a 127b 140
Mr. P. Moss   99a
Mr. E. Paget-Tomlinson   151b
Miss M.A. Parsons   119b
Mr. and Mrs. A. Picken   109b 124 134b 139a
Mr. C. Priday   128
Royal Commission on Historical Monuments (England)   83 109a 115b 116 117 118 119a 129b 137a 139b 141b 163
Mr. B.L.H. Shaw   107a 133b
Mr. A.G. Simmonds   93 149a
Mr. A. Thomas   108 138 145a 145b
Mr. M.E. Ware   106 121a 134a
Mr. H.I. Williams   114

Introduction

Gloucester was given the formal status of a port by a charter from Queen Elizabeth in 1580, but few sea-going ships ventured up to the city quay because of the difficulty of navigating the tidal and treacherous River Severn. Some cargoes were unloaded at Newnham and Gatcombe, but Bristol remained the principal port in the Severn estuary. By the eighteenth century, a very extensive traffic had built up between Bristol and the Midlands, with a wide range of cargoes being carried in shallow draught sailing vessels known as trows (pronounced to rhyme with crows). During the second half of the eighteenth century. Newnham developed as a transhipment port for goods being carried round the coast to and from London, and a group of local merchants operated several brigs on this service. Also, a number of canals were built to link up with the Severn, including the Staffordshire and Worcestershire Canal opened in 1772 and the Stroudwater Canal opened in 1779. These developments encouraged a continuing growth in traffic on the river, and this highlighted the restriction caused by the narrow winding stretch below Gloucester which could only be navigated for a few days each month on the spring tides.

To bypass this restriction, a proposal was made in 1783 to construct a canal from the Severn at Gloucester to join the Stroudwater Canal near Whitminster. This was linked with the plan for the Thames and Severn Canal which would then give an inland waterway through route to London. The Gloucester Canal Committee wanted to form a basin just to the south of the existing riverside quay, but they found that the new county gaol was due to be built there, and after various unsuccessful attempts to get the gaol scheme changed, this first canal proposal was not persued.[1] A few years later, however, a group of merchants and bankers from Gloucester and other towns up the river came together to promote a far more ambitious scheme. The new proposal was to build a canal between Gloucester and Berkeley Pill that would be large enough to admit sea-going ships of three or four hundred tons burden. This would not only allow the existing through traffic to avoid the worst part of the river, but it would also help to develop the port of Gloucester as a rival to Bristol.[2] An initial meeting was held on 6 November 1792, and with canal mania gripping the country, there was no difficulty in attracting subscriptions.[3] After some rather hurried surveys of the route, an Act of Parliament was passed in March 1793 authorising the construction of the canal and the formation of a company with the power to raise up to £200,000.[4]

The subsequent activities of the Gloucester and Berkeley Canal Company were controlled by a twice yearly general meeting of proprietors with other special gatherings if circumstances required. At the regular meetings, the proprietors elected a committee to run the day-to-day business for the following six months. In the early years the committee was mainly drawn from among the leading citizens of Gloucester, and the principal members were Thomas Weaver, a pin manufacturer, Richard Chandler, a woolstapler, Richard Brown Cheston, a doctor, Charles Brandon Trye, a surgeon, Thomas Mee Esq., William Fendall, a barrister and banker, Edwin Jeynes, a mercer and banker and Giles Greenaway Esq. These men carried a heavy responsibility as they were embarking on one of the biggest civil engineering projects so far undertaken. They appointed Robert Mylne as chief and principal engineer, and he carried out further surveys of the route and produced detailed plans and sections. Mylne was one of the leading architect-engineers of the eighteenth century who had made his name by designing Blackfriars Bridge across the Thames. Although now sixty, he was still very active designing buildings and advising on river and canal schemes, including an attempt to improve the River Severn.[5] Full of hope for an early link with the Stroudwater Canal, it was agreed that work would start at Gloucester where a major task would be the excavation of the great basin that was destined to become the nucleus of the present dock complex.

The proprietors soon found that they had taken on a more difficult task than they had expected. The main problem was rising costs due to the effects of the Napoleonic War, but matters were not helped by small inefficient contractors, poor supervision and appalling weather conditions. These difficulties eventually proved too much for the original proprietors, and it was left to a later generation to complete the massive project. The final cost was more than twice the originally authorised capital, and much of the additional money had to be raised by loans which took over forty years to repay. This burden of debt was a great handicap to the financial prosperity of the Canal Company, but it did not prevent the canal and the docks making an important contribution to the commercial life of the city. The original proprietors may have lost all their investment, but the citizens of Gloucester were to have good reason to be grateful to them.

The following chapters tell the story of the early struggles to build the

A trading token halfpenny issued for local use in 1797 during a scarcity of small change.

dock basin at Gloucester and of the subsequent developments to accommodate the growth in trade during the nineteenth century. The last two chapters summarise the great changes that have taken place in the pattern of trade during the twentieth century and briefly describe the condition of the docks in 1983. As the warehouses remain such a striking aspect of the scene, additional information on their design and construction is included in an appendix. Other appendices give a summary of the more important dates associated with the construction of the docks, some indication of the annual tonnages handled, and a list of the principal sources that have been consulted. The list includes the specific references numbered in the text and also identifies the volumes of the Canal Company's records used as the basis for each of the chapters.

A gang of men towing a vessel up the narrow winding stretch of river approaching Gloucester *circa* 1740. The basin of the Gloucester and Berkeley Canal was later formed in the meadows to the right of the picture, and the island which the vessel is passing became the west bank of the basin.

CHAPTER ONE

Great Excavations

The site chosen for the basin was on the southern edge of Gloucester in meadows by the River Severn. The plan was for the ship canal to enter from the south and for a double lock to the north to provide access to the river for trows and barges. There was an island in the river at this point known as the Ait, and part of the channel to the east of it was chosen as the site for the lock. To prepare for the main excavation, therefore, this channel was blocked off in the summer of 1794, and a former house on the island was adapted to provide a workshop. By the autumn, the land for the basin itself and for the first part of the canal had been purchased, and in October the Canal Company started advertising for canal cutters. To help potential contractors appreciate the task ahead, arrangements were made to dig out the first 450 yards of canal to provide a specimen of what was expected, and a model was prepared of the basin and the lock as they were thought to be 'rather of an intricate and unusual nature for most mens comprehension'. The original plan was that the basin should be hexagonal in shape, but the chief engineer, Robert Mylne, recommended that only the western half should be dug initially, leaving the eastern half to be financed later by the expected profits from trade. Dennis Edson was appointed resident engineer to supervise the day-to-day work of the contractors, as Mylne was only expected to visit the works periodically because of his many other commitments.

The first contractor for the basin was Sampson Lockstone of Sodbury, and he agreed to dig down to a depth of six feet or to the brick earth if found sooner. At the end of October, there were fifty workmen employed and each was given one shilling for drinks to celebrate the start of work. Within a few weeks, however, operations were disrupted by flooding. A steam engine for pumping had been supplied by Boulton and Watt, but there had not been time to erect it and a supply of hand-operated pumps was hastily arranged. The Canal Company evidently accepted some responsibility for the situation as Lockstone was compensated for having to get the water out of his workings and Edson was ordered to divert an open sewer that had probably been the cause of the problem. There then followed a period of particularly severe weather during which the River Severn was frozen over for almost a month. When the thaw came, it was accompanied by heavy rain, and the swollen river burst its banks in many places.[1] It is not clear if the excavation was flooded this time, but the appalling weather certainly prevented any real progress being made for four months. To make matters worse, Lockstone went down with consumption and had to give up his contract, and when the committee inspected the site, they were dismayed to find that the banks were slipping in several places and at high tide the river water oozed through into the workings.

The committee quickly arranged for Thomas Baker of Eastington to take over the work, and he also took over the planks, wheelbarrows and ladders left behind by Lockstone. The plan was to divide the excavation into four parts leaving narrow dams between, and the Company were to find pumps and labourers to empty water from one division into another so that one was always dry enough to work in. With the previous winters experience fresh in their minds, the committee agreed that the banks of the basin should be raised five feet above the water line to exclude any flooding from the river, that four larger pumps should be ordered, and that a nearby drainage ditch should be prepared to receive water pumped from the workings when the water in the river was high. As the weather improved, a landing stage was built on the river bank for unloading all kinds of materials, and a lime kiln was constructed nearby. It appears that the excavation of the basin relied entirely on manual labour, with gangs of navvies loading the earth into barrows and wheeling it over lines of planks to where it was to be dumped. Some of the soil was used to build up the bank by the river to guard against further floods, but much was dumped to the east of the excavation with little thought being given to the effect this would have on completing the other half of the basin in the future. As the workings went deeper, water became a problem again, and 5,000 gallons had to be pumped out each day. To meet this situation, the steam engine was set up on the bank between the workings and the river, and drains were dug to feed the water down to a pit from which the engine could pump it into the river. In initial operation, the performance of the engine fell far short of its specification, but at a special trial attended by a representative from the manufacturers, the engine did achieve its full output 'with some difficulty'.

In May 1795, contracts for bricklaying and stonework were agreed with George Stroud and Daniel Spencer of Gloucester and Thomas Cook of Brimscombe. Cook had worked as a master builder on the Thames and Severn Canal and had also built the county gaol at Gloucester. The bricks were expected to be made on the site, and transport of two thousand tons of stone from Brimscombe was arranged

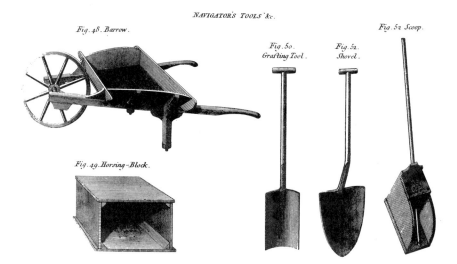

Fig. 48. Barrow.

Fig. 52. Scoop.

Fig. 50. Grafting Tool.

Fig. 51. Shovel.

Fig. 49. Horsing-Block.

Tools of the type used by the navvies excavating the basin. Horsing-blocks were used for supporting the planks along which the wheelbarrows were pushed to remove the soil.

with Humphrey Brown, a carrier on the Severn. Good progress was made in excavating the upper part of the basin, but Thomas Baker was forced to give up his contract in July due to rising costs caused by wartime conditions. The work was taken over by George Mills from Staffordshire, and he was also given the contract for the remaining lower part of the basin, but this was suspended when James Brassington offered to dig out the clay and make it into bricks for only a small additional cost. Brassington started work but it was soon found that the bricks he could make were not fit for use, and an alternative supply was obtained from the Herefordshire and Gloucestershire Canal Company. Brassington continued digging for a few months, but with winter approaching, he was ordered to build up the banks against possible floods and to improve the drainage down to the engine pit and then he too gave up the work. By this time the committee had become thoroughly dissatisfied with Dennis Edson, and he was replaced as resident engineer by James Dadford, whose family included canal engineers but who had little relevant experience himself.

In November 1795, Charles Holland became the fifth contractor in little over a year to take on the awesome task of excavating the basin. As a further precaution against floods, he was first asked to dig out the south west corner of the basin into a square shape and to deposit the soil so as to reinforce the bank along the western side of the canal workings to the south. Then in January, Holland started on the remaining part of the basin that was due to be done, and eventually went on to finish this. Thus the basin became almost rectangular in plan and only the angle of the north end remained to indicate the original intention to make it hexagonal.

With the excavation work going well, attention turned to the building of quay walls along the north and west sides of the basin. The east side was to be left as an earth bank as there was still hope of extending in this direction eventually in spite of the piles of soil that had been dumped there. A small ceremony to celebrate the laying of the first stone was attended by Mylne on 8 April 1796,[2] but all was not well. A number of proprietors were getting concerned about rising costs and about some aspects of Mylne's designs, and two eminent engineers Robert Whitworth and William Jessop were asked for their views. In particular Charles Brandon Trye expressed concern that the quay walls were to be built upon a bench of clay about nine feet above the bottom of the basin. Whitworth noted that although the clay bench appeared to be hard, it was likely to become soft when under water, and he pointed out that wooden stages would need to be built out from the quay wall for large boats to come alongside and that this would restrict the waterspace. He recommended instead building full height quay walls in limestone with a few ashlar courses which would be cheaper than the proposed half-height walls all in ashlar. In view of this report, wall building was stopped temporarily, but at the next meeting of proprietors a resolution of no confidence in Mr. Mylne was narrowly defeated and the wall building went ahead again. Time was to show the folly of this decision.

During the following year, work continued under Dadford's supervision with little involvement from Mylne, who was having to cope with family problems as well as his other professional commitments. This led Trye to complain again, and in June 1797 he wrote to his fellow committee members suggesting that Mylne should be paid by the day rather than by an annual salary. He commented that he did not suppose that any member of the committee was now amused by 'the idea, which in the days of our ignorance was broached, that an engineer may render us sufficient services by thinking and contriving for us while sitting at his ease in London'. The committee initially deferred any action on this letter, but with concern growing about rising costs, they wrote to Mylne asking for his opinion about omitting the transom pieces and cross planking from the floor of the double lock that would give access to the river. Two weeks later, they had had no reply from Mylne, and so they accepted Dadford's advice that the woodwork was not really necessary and authorised him to complete the lock according to his discretion. When Mylne did eventually come to Gloucester in the autumn, it seems that there was a row about the change in design, but the committee over-ruled Mylne's objections. They also pressed the matter of his salary and he was forced to accept payment by the day as Trye had proposed.

The change in the design of the lock was soon to feature in a further row when the north quay wall collapsed during the very severe winter of 1797-8. Mylne blamed this on poor workmanship and on the effects of water that had been let into the basin while the river was in flood. The water had entered through a hole in the bank that had been made for laying a new culvert, and the basin had been filled almost to its intended level. At the time, the *Gloucester Journal* commented that 'the accident fortunately occasioned little inconvenience or injury and gave the town an opportunity of forming a judgement of the handsome appearance the undertaking is destined to make'.[3] After the collapse of the wall, however, Mylne commented that the water had been allowed to get behind the wall at its western end where the lock was still incomplete, and he criticised the committee for giving authority for the change in design. Dadford strongly disagreed, claiming that the collapse was due to the winter weather acting on the bench of clay on which the wall was built. Mylne defended his design by saying that the west wall was still standing, but Dadford pointed out that in some sections the foundations of this wall went down to the bottom of the basin (as the clay bench had not been found) and in other places where Mylne's design had been followed closely there were in fact signs of collapse. Whatever the truth of the matter, at a General Meeting in March 1798, the proprietors backed Dadford and they resolved that Mr. Mylne be no longer employed. In hindsight, this action was strongly criticised by a later engineer, who thought that Dadford was quite unfit for the work and was 'as much inferior to Mylne as a glow-worm is to the sun'.[4]

Work then continued under the authority of Dadford alone. The signs of collapse in the west wall were attended to, and the north wall was taken down and rebuilt on a more substantial plan. By March 1799, the lock was finished and the basin was ready for use. The digging of the canal, however, had only progressed south as far as Hardwicke, about five miles from Gloucester, and most of the original share capital had been spent. An attempt to raise further capital had not been successful, and it was not thought proper to use the remaining money to extend the cutting of the canal but rather just to finish off what had been started.

Local boat owners who were trading from the old city quay by the river asked if they could start using the basin for a reasonable charge, but the committee did not think that they had the right to charge tolls less than those laid down in their Act of Parliament, and so no arrangement could be agreed. Thus the basin and that part of the canal that had been completed were left full of water while the proprietors considered what to do next. With no further construction work in progress, James Dadford's services as resident engineer were dispensed with.

For the next few years, there was much discussion about how the canal could be completed at reasonable cost and how the necessary money could be raised. The committee considered various schemes for linking up with the Stroudwater Canal or for taking a short route to the River Severn at Hock Crib near Fretherne, but it was still not possible to raise the necessary money. Then in 1806, a new matter was brought to the attention of the committee. Plans were being prepared for a tramroad between Cheltenham and Gloucester communicating with the river by means of the basin. The tramroad was an early form of horse-operated railway with flanged cast-iron rails mounted directly on to stone blocks. A similar tramroad was being planned to bring Forest of Dean coal down to the River Severn at Bullo Pill near Newnham, and the intention was for coal to be carried by boat to Gloucester and then by tramroad again to Cheltenham. The route eventually chosen for the tramroad started by the city quay, passed near the basin and round the southern outskirts of Gloucester, and then turned north to pick up the line of the turnpike road to Cheltenham. A branch connected with Charles Brandon Trye's quarries on Leckhampton Hill.[5]

The Canal Company strongly supported the project, but at first did nothing about opening the basin. Instead, the idea arose of forming a wharf on the river bank by the south-west corner of the basin which would be reached by a temporary bridge over the canal. The Bullo Pill company tried to get an exclusive lease of this wharf, but the Lydney and Lydbrook (later the Severn and Wye) tramroad and local boat owners asked for equal rights. The Canal Company therefore agreed to allow anyone to use the wharf, but they thought that they would raise some revenue by charging a toll for goods carried over their land, not realising that they had no powers to do this. The tramroad was opened throughout its length on 4 June 1811 although some traffic had started earlier.[6] The tramroad company did not act as operators, but allowed anybody to use their own wagons and horses on the rails for payment of a toll. To try to collect their additional toll, the Canal Company erected a pair of gates across the rails, but this brought an immediate protest from the tramroad company, and it was eventually agreed instead that

18

The design for the seal of the Gloucester and Cheltenham Railway Company, symbolising how the tramroad linked the Leckhampton quarries with the River Severn.

the latter would pay a rent for the land on which the rails were laid. The Canal Company considered improving the river wharf by building jetty heads with cranes or sheers for which a toll could be charged, but it is not clear if this was ever carried out. It seems more likely that the combined effects of the current and the tides made use of the river bank very difficult, and in June 1812, the Canal Company reversed their earlier policy and decided to open the basin.[7] Arrangements were therefore made to clean out the mud that had been allowed to accumulate in the lock, and the basin was opened in October 1812.[8]

The basin was soon in regular use by trows bringing Forest of Dean coal from Lydney and Bullo Pill, and also by boats bringing timber and roadstone from further down the estuary. This usage showed up a few problems, and the job of sorting them out was given to John Upton who was appointed clerk to the Canal Company in 1813, and also took on the job of engineer. The main problem was the lack of a proper water supply as the canal had not been completed far enough to collect the intended main feed from the River Cam. The problem was made worse by leaks in the canal and by faulty construction of the lock which would empty itself in two hours. Upton was told that the gate cills had been merely bedded in mortar on the natural clay, and he recommended that stout grooved piles should be well driven down in front of the cills. In the summer of 1814, the combined effects of evaporation, leakage and usage of the lock caused the water level in the basin to drop four feet. Upton set about stopping the leaks and cleaning and enlarging what feeders there were, and arrangements were made to re-erect the company's steam engine to pump water from the Severn. The other main problem was that of access, and Upton made a new road link between the basin and the city and also laid several tramroad sidings on the east side of the basin.

The principal tenants on the Canal Company's land were William Prosser and Robert Lovesey who had timber yards at the north end of the basin, and Samuel Playne who had a long established rope walk to the east of the basin. Other traders apparently just used the public wharfs for transhipping goods. A separate development on private land just to the south of the basin was the establishment of a boat-building yard by John Powell. The *John Guise* (113 tons register) was launched in 1814, and was probably the first deep sea vessel to be built in Gloucester. It was a great day when she departed on her maiden voyage and she was seen off by hundreds of cheering spectators.[9] Her owner Thomas Smith tried to avoid paying the full toll for building and launching her on the grounds that he had already paid toll on much of the timber in her, but the Canal Company would not agree to any reduction. She subsequently returned to Gloucester many times, often bringing timber from the Baltic or Scandanavia.

Some of the stone blocks from a siding of the Gloucester and Cheltenham tramroad uncovered during an excavation in 1983.

CHAPTER TWO

Completing the Canal

With the basin now in use, attention concentrated on how to complete the canal. For several years, serious consideration had been given to a shorter line of canal joining the Severn at Hock Crib near Fretherne, but it had not been possible to raise enough money even for this. In 1815, John Upton wrote a paper attacking both the original line to Berkeley Pill and the line to Hock Crib, and proposed instead a new line joining the Severn at Sharpness Point.[1] These observations created a considerable stir among the proprietors, but he was supported by Mark Pearman, an enthusiastic new shareholder who had taken an interest in the canal while on holiday in Gloucester, and in the following year the plan was formally adopted at a special assembly of the proprietors.[2] The problem of raising finance was eased by the promise of a loan from the Exchequer Bill Loan Commissioners who had been set up by the Government to help relieve unemployment after the Napoleonic War. Some work started in 1817, and in the following year the project came under the general supervision of Thomas Telford acting as consultant engineer to the Loan Commissioners. Upton initially acted as resident engineer, but he was accused of irregular dealings and was replaced by John Woodhouse in 1818. Construction work progressed steadily, and the junction with the Stroudwater Canal was made on the 28 February 1820.[3] This completed an inland waterways route between the Midlands and London, and more than twenty vessels passed along the new line on the following day.

In anticipation of more boats using the basin, arrangements had been made with John Bird of Stourport to provide proper repair facilities. In 1818, Bird was given a fourteen year lease of land beside the south-west corner of the basin for him to build a graving dock one hundred feet long by thirty-four feet wide and at least ten feet deep. He was allowed to use water from the basin to fill the dock and to use an existing culvert to the river for draining the dock. As well as carrying out repairs, he went on to build a number of trows. As more boats started using the basin, it became increasingly difficult to collect the tolls for goods stored on the wharfs, and the committee decided to divide up the adjoining land into separate yards that could be let to the traders. Those who took the larger yards included Thomas Harding and Sinclair

Hendrie, who were coal merchants in Gloucester, and John Roberts who was the Cheltenham agent of the Bullo Pill Coal Company. Several of the other yards were also occupied by coal merchants, two were taken by stone masons and one by a china merchant.

Some work continued on completing the canal to Sharpness, but it was delayed due to difficulties in raising extra money, and within a few months it stopped altogether when the contractor went bankrupt. After lengthy negotiations, the Loan Commissioners agreed to lend more money, but they effectively took control of the company. The final stage of constructing the canal then began in 1823, with Hugh McIntosh, an experienced contractor from London, working under the supervision of Thomas Fletcher who had been appointed resident engineer in 1820 on Telford's recommendation. Fletcher reported regularly to Telford, who continued to visit the works periodically and who sent reports and recommendations to the Loan Commissioners. To represent the interests of the proprietors, George Nicholls was appointed permanent chairman of the committee with full authority to act in its name, and he spent a considerable amount of time superintending the company's affairs. These parties kept in close contact by meetings and correspondence, and the vast project was soon making good progress again.[4]

With the completion of the canal now in prospect, doubts arose about whether the basin at Gloucester was big enough. It had only been built to half the originally intended size, and it had quay walls only along the west and the north sides. As the water was due to be drained off during 1824 to allow the accumulated silt to be removed, it was decided to use the opportunity to build a quay wall along the east side of the basin and to construct a separate Barge Basin to the south-east. Telford recommended that the new quay should be four feet seven inches above the water with wooden defenders to suit sea-going vessels and that the Barge Basin should be ten feet deep with two foot high walls to suit river and canal boats. The soil excavated could be dumped on some low ground to the south. Work started in the spring of 1824, and was expected to take about six months. Draining off the water proved difficult, however, as the bottom of the basin was lower than the river. There was further delay during August when many of the men left to help bring in the harvest. Then heavy rain in October flooded the works and threw everything into confusion, and so McIntosh promised to bring his steam engine from Tewkesbury and set it up by the lock at Gloucester.[5] It was also found that the amount of mud that needed to be removed from the old part of the canal was much greater than had been thought, and it was about a year before the water was re-admitted. This caused considerable disruption to the commercial life that had de-

Trows using the Basin *circa* 1823 before the canal was open for seagoing ships.

veloped round the basin, and the Canal Company agreed to a reduction in rents for their tenants by way of compensation. The Barge Basin was completed in 1825, and the adjoining land was divided into nine yards on either side, each yard being served by a siding connected with the Gloucester and Cheltenham tramroad.

Also in 1825, the Canal Company seriously turned their attention to the warehouses that would be needed at Gloucester to store the goods that they hoped would soon be carried on the canal. As they were so heavily in debt to the Exchequer Bill Loan Commissioners, they decided to sell off some land for building private warehouses and to put the proceeds towards a modest warehouse for themselves. They first asked Telford to 'furnish a plan for wharfs and warehouses with proper approaches thereto', and he recommended that the Company should build two moderate size warehouses to the north of the Main Basin. Then a member of the committee, John Gladstone, agreed to arrange for a competent person to provide detailed plans. Gladstone was a prominent Liverpool merchant who had only come to Gloucester so that his daughter could benefit from the medicinal waters of the Spa.[6] He contacted a Liverpool builder, Bartin Haigh, who then prepared plans for two types of warehouses and provided a layout for the land around the Main Basin. Haigh's scheme showed three semi-detached warehouses of five storeys to the north of the basin and four huge blocks each of five large warehouses to the east of the basin. The whole scheme was rather grand for such an early stage in the development of the docks and was presumably meant to indicate long term intentions. It certainly appealed to the proprietors, and Haigh's plans and specifications received 'unqualified approbation' at a special meeting on 30 December 1825. It was agreed that initially one block of land to the east of the basin should be sold off in five lots for building private warehouses, and that the Company should build two of the smaller warehouses to the north of the basin as soon as money was available.

The excitement felt at that meeting was to be short lived. 1825 had been a year of financial crisis for the country as a whole, and many people had lost money as banks had failed. The prospect of building warehouses with no guarantee of any goods to fill them was evidently not attractive to private investors, and two months later not one offer had been received for any of the plots. This set-back put in jeopardy the Company's plans to build their own warehouses, and the committee realised that they had to take urgent action as the canal was due for completion in four months and some form of warehousing was essential. The chairman was dispatched to London to see if the Loan Commissioners would provide some more money, and a letter was sent to Haigh asking if there was any objection to building the warehouses to only three storeys initially and adding on the other two storeys later if required. Haigh recommended building to the full height for an estimated cost of £6,380, and the Loan Commissioners agreed to back the scheme as they accepted the need. The committee therefore advertised for tenders, and such was the urgency, that they allowed only three weeks for the tenders to be submitted and asked for the building to be completed in six months.

William Rees and his son William of Barton Street put in a tender of £6,600 for the two semi-detached warehouses, and this was accepted subject to the provision of adequate sureties. After a slight delay, Rees got two friends to guarantee £1,000 in case of any default, and they also

The North Warehouse built by the Canal Company as two semi-detached units and completed just in time for the opening of the canal in 1827.

became parties to the contract. These friends were George Williams, who had a timber yard by the basin and would be supplying the timber for the warehouses, and John Chesterton who ran the Talbot Inn in Southgate Street. The contract specified that the bricks were to come from 'up the river above Westgate Bridge', the timber was to be from Memel or Dantzic, the stone from Bath or the Forest and the slates were to be 'large blue Welsh ton slates'.[7] The contract was signed on 18 May 1826 and the warehouses were to be ready for occupation by 18 February 1827. (The earlier urgency had relaxed as it was now acknowledged that the completion of the canal would be delayed). Construction was well under way when the Canal Company decided to reduce the height from five to four storeys probably because the subsoil was found to be unfavourable. They negotiated a reduction in price of £820 with William Rees, but George Williams was not party to this and subsequently claimed £25 compensation for providing more timber than would now be needed. The building, which later became known as the North Warehouse, was completed in April 1827 just in time for the opening of the canal. The Canal Company initially objected to the hanging of the window shutters, but when Telford was brought in to adjudicate, he ruled that they were sufficiently near the description in the specification to be acceptable. The building was designed so that individual floors in each part could be rented to different merchants who had common access to the roof space to operate the hoists over the loading doors. The brick vaulted cellars were thought to be suitable for storing wines and spirits, but it was found that the doors and windows did not meet the standard of security required by the Customs for bonded stores and so modifications were made before the cellars were used.

While the warehouse building had been going on at Gloucester, good progress had been made in completing the canal, and the committee had considered the arrangements needed for eventual operation. They identified the employees who would be needed, they proposed bye-laws and they drew up a table of tolls. They recognised the need to have a qualified engineer in charge, and William Clegram from Shoreham was appointed harbour master, engineer and general superintendent although Fletcher was retained to supervise the completion of the canal. The opening was fixed for 26 April 1827, thirty-four years after the original Act of Parliament had authorised the start of work, and arrangements were made to ensure that two vessels would be available to enter the canal on that day. The first to pass through the lock at Sharpness was the schooner *Meredith*, a locally owned boat with a cargo of brandy from Charente (France). Of more interest, however, was the much larger ship *Ann* of 300 tons register which had come from Bristol to pick up a cargo of salt for the Newfoundland fisheries. Once through the lock, both vessels hoisted their colours and manned their tops, and with the help of towing horses they set out on their historic journey along the canal. Boats carrying members of the committee and a band were waiting at the junction with the Stroudwater Canal, and a number of pleasure boats also joined the convoy. An ever increasing number of spectators walked along the towpath until the crowd lining the banks became almost too dense to move. Eventually the vessels entered the basin at Gloucester amid the firing of guns, the ringing of church bells and the cheering of the large mass of people who crowded every vantage point to get a better view. The sight of a full-rigged ship, decorated with flags and streamers, coming to her moorings in the heart of the city generated much admiration and excitement. Tents and booths with liquor and refreshment lined the margin of the basin, and the *Gloucester Journal* reported that 'it was long ere the novelty of the scene would allow the collected assemblage of spectators to disperse'.[8] For the proprietors of the Canal Company and their friends there was a dinner at the Kings Head, and over twenty toasts were drunk. In view of the strained financial circumstances of the Company, however, it was agreed that the expenses of this had to be borne by those who attended, but free supper and beer were provided for the workmen who had been employed on the mammoth enterprise that was now successfully completed.

With the canal now fully operational, activity at the basin increased, but initially most of the boats were just small river and coastal traders using the canal to bypass the winding tidal stretch of the river below Gloucester. The appearance of a large vessel aroused great interest, and when the brig *Alchymist* arrived in August 1827 with 360 tons of deals and battens from Archangel, many people walked down the canal to meet her and church bells were rung when she tied up in the basin.[9] Gradually traffic began to pick up as local merchants started importing goods from Ireland and the continent, and then merchants from other towns also came to take advantage of the new facilities. The advantage compared with Bristol was that cargoes could be transferred directly from sea-going ships into canal longboats (often called narrow boats in other parts of the country). These could carry the goods throughout the inland canal system to supply the growing industrial towns of the Midlands. With Bristol also having high port charges at this time, traffic through Gloucester was soon exceeding all expectations.

One of the early trades to develop was the importing of corn directly from Ireland, instead of it being transhipped at Bristol. With no previous experience of corn handling to go on, the local merchants initially agreed to pay a rather high rate per bushel for unloading, but when they found that the men could earn a normal weeks wages in only two days, the rate was quickly reduced.[10] Wheat, oats and barley were brought over in small boats from places such as Limerick, Wexford, Waterford and Dublin. There were also occasional shipments of corn from Hamburg and other continental ports, but the number of these fluctuated considerably depending on the quantity of the home harvest. Before 1827, the import of foreign corn had been prohibited unless the average price at home exceeded a given level. The intention was to protect British farmers, but the effect was to cause much hardship for the poor in years of bad harvest. To try to stabilise prices, the Corn Law of 1828 laid down sliding scales of import duties related to current market prices, and this led to an increase in foreign imports. The opening of the canal allowed Gloucester to capture a significant share in this new trade which was to become of great importance later in the century. Another early trade that was to become of great importance was the importing of timber, and ship-loads of rough-cut logs and sawn deals and battens were soon arriving from the Baltic, Archangel and

Canada. Before the canal was open, local merchants had arranged for cargoes to be off-loaded down river and brought up to Gloucester in barges or in the form of huge rafts towed by gangs of men struggling their way along the river bank. With the opening of the canal, however, the ships could come right up to the basin to unload, and the historian Counsel noted that 'the demand for timber has increased to a prodigious extent' and it could be bought in Birmingham cheaper than at Bristol or Liverpool.[11]

Another existing trade was the importing of wine, fruit and cork from Portugal. Since the 1790s, some local merchants had risked bringing their boats up to the river quay at Gloucester at high spring tides, but delays and mishaps were common and the opening of the canal proved a tremendous benefit. Other imports included brandy from France and barilla (for soap making) from Teneriffe.[12] Unfortunately there was a shortage of return cargoes, and many ships had to leave in ballast. The only regular export was salt from Droitwich and Stoke Prior in Worcestershire. This was brought down the Severn in barges and long boats and transhipped at Gloucester for dispatch to Ireland and occasionally to Europe. In the coastal trade, the main inward cargoes were slates from North Wales, copper and zinc from South Wales and wool from Bridgwater, with salt again sometimes taken as a return cargo. Many thousands of tons of slates were sent on to supply Birmingham until a toll reduction on the northern canals made the route via Chester and Runcorn cheaper again. There were also regular traders to Bristol with general cargoes, and there were many small boats bringing coal from the Forest of Dean and stone from the Avon Gorge.

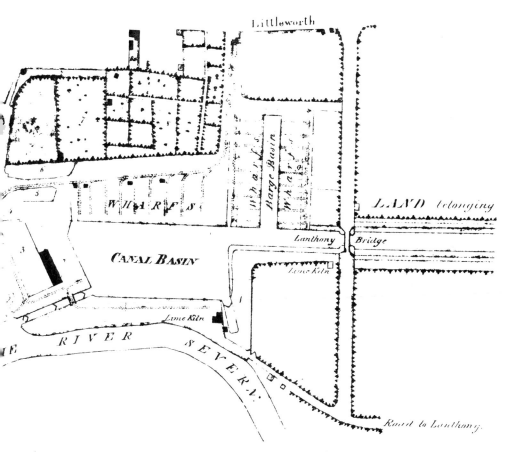

Sutherland's map dated 1829 – soon after the canal was opened. The Canal Company's warehouse is shown near to the lock, with a timber yard (3) and the original lock keeper's cottage (4) nearby. The graving dock is in the south-west corner of the Main Basin, and there are many tramroad sidings serving the wharfs along the East Quay and around the Barge Basin.

CHAPTER THREE

Providing Additional Facilities

As trade developed, the demand for storage space soon exceeded that available in the Canal Company's warehouses and arrangements began to be made for building private warehouses along the west side of the basin. At first, the Canal Company was unwilling to grant long term leases for the land, and Humphrey Brown and Son (an old established firm of river carriers based at Tewkesbury) started by building just a single storey warehouse not far from the graving dock. Under pressure from the Exchequer Bill Loan Commissioners, however, the Canal Company then adopted a standard sixty three year lease, and this encouraged the building of more substantial warehouses. The first to avail themselves of the new arrangements were Joseph and Charles Sturge, corn merchants of Birmingham, who put forward plans for buildings with three storeys and cellars. These were immediately approved by the Canal Company, and with a commendable consideration for appearance as well as utility, the committee decided that they wanted all other warehouses to conform to the same general style. When Humphrey Brown applied for land to build two further warehouses, therefore, agreement was only given on the condition that he would also pull down his single storey warehouse and replace it with a three storey building. Thus it was that a 380 foot long row of similar warehouses (now demolished) came to be built along the West Quay during 1829-30 (with Brown's rebuilding probably completed a year later). Messrs. J. & C. Sturge had three blocks in the middle of the row, Humphrey Brown and Son had three blocks at the south end and the Droitwich Salt Company had a double block at the north end. Several of these warehouses were registered with the Customs as bonded warehouses where corn and other imports could be stored without paying duty, and the cellars were usually let separately to wine and spirit importers.

During the same period, other developments were also taking place. In 1830, a warehouse was built for John Biddle near the corner of the Barge Basin. Biddle was a prominent miller from Stroud who imported corn through Gloucester and then shipped it in barges to his mills via the Gloucester and Berkeley and the Stroudwater canals. Also in 1830, the Canal Company decided to build an office at the north end of the

The northern end of the row of warehouses along the West Quay shortly before demolition in the 1960s. The influence of the Canal Company's policy of requiring a reasonable uniformity of structure and appearance is evident.

Main Basin to provide a proper base for the administration of their expanding business. The building was completed in the following year and included living accommodation for W.B. Clegram, the son of the Company's engineer, who had been appointed clerk in 1829. This father and son partnership was to give loyal and valued service for over thirty years.

Apart from Biddles Warehouse, most of the east side of the Main Basin was devoted to the timber trade. The old established local firm of Price and Washbourne had yards at the north end and Maurice and James Shipton of Birmingham had a large yard by Biddles Warehouse. Also, Evan Jones had a bonded yard beside Llanthony Road where foreign timber could be stored without paying import duty. In 1830, the Canal Company formed another bonded yard immediately to the north of their North Warehouse, and this was leased to Benjamin Tripp. To meet the standard of security required by the Customs, this yard was surrounded on three sides by a high brick wall with a pair of large wooden gates, and the Canal Company agreed to insert iron bars in the lower windows of the North Warehouse to prevent any access to the yard from that side. Much of the wood arriving at these yards was already shaped into deals and battens, but there were also many large balks that needed to be cut up by pairs of sawyers using large double-handled saws worked by one man on top of the log and another man in a pit beneath. Wood offered for sale included fir from the Baltic, red and yellow pine from Canada and mahogany from Honduras, and most timber merchants also stocked various sizes of slates that had been brought round the coast from North Wales.[1]

Two other yards on the east side of the Main Basin were occupied by William Kendall and George Ames, who were wharfingers and ship-brokers handling a wide range of general cargoes. They organised several boats a year trading between London and Gloucester, although the timings were somewhat irregular as the journey could take anything from one week to three months depending on the weather.[2] Of the yards around the Barge Basin, eight were occupied by Thomas Harding, a coal merchant and stone mason. In 1830, the committee noted that he was only paying toll on the canal for business that could be handled by two yards, and they gave him notice to quit six of the yards to make way for people who would bring the Company more income. The other yards were mainly used by other coal and timber merchants with changes in tenancy occurring quite frequently. A surviving inventory shows that one yard occupied by Humphrey Brown was surrounded by seven foot high fencing with a pair of gates which opened on wheels. By the quay was a manually operated cast-iron crane capable of lifting seven tons, and in the yard was a tramroad wagon and several wheelbarrows presumably used for transferring cargoes between the boats and the tramroad.[3]

On the opposite side of the canal to the Barge Basin, the ship building yard had been re-established by William Hunt. He launched some small schooners there, but when he planned to build a vessel of one hundred and thirty tons register in 1831, the Canal Company became concerned about the effect that the launching would have on the works of the canal and he agreed to move to another site near the river. It seems that the experience gained with this must have been re-assuring as he was later allowed to launch even bigger vessels into the canal.

Traffic on the canal had increased rapidly, and in the early 1830s, over seven thousand boats a year were paying dues on over 300,000 tons of cargo.[4] Most were canal longboats or trows and barges employed in the river and coastal trade, but there were also increasing numbers of two-masted brigs and schooners and three-masted barques, some carrying over 400 tons of cargo. In February 1832, Telford noted fifty-eight vessels in the basin of which thirty-three were classed as sea-going.[5] A few months later, the *Gloucester Journal* reported with satisfaction that there were eleven large boats unloading timber from Canada, Archangel and the Baltic, six boats with corn from Ireland, one with brandy from France, one with wool from Hamburg and one with a

A sketch of the original engine house based on contemporary illustrations.

general cargo from London.[6] All this activity put the available facilities under great strain and highlighted the need for further developments. Every action of the Canal Company, however, was still subject to the approval of the Exchequer Bill Loan Commissioners, who wanted all surplus revenue to go towards repaying their loan. Much pleading was required to get any new expenditure authorised, and many letters were to be exchanged over the next few years between the Canal Company and the Commissioners in London.

A particular problem was the provision of an additional source of water for the canal, as the supply available from the River Cam and other streams was proving inadequate. After lengthy negotiations with the mill owners in the Stroud valley and a special Act of Parliament, arrangements were made to take some water from the River Frome. It was also agreed to provide a steam engine and pump at Gloucester. A forty-five horse-power beam engine was ordered from Graham and Company of the Milton Iron Works in Yorkshire, an engine house was built beside the graving dock with a culvert from the river, and the engine was erected and set to work in 1834. Another problem was the graving dock itself. This had only been built for repairing the small vessels trading in the river before the opening of the canal, and it now needed enlargement and renovation. Initially it was agreed that John Bird could remain as tenant on condition that he put the dock into good repair, but it seems that the improvements were going rather slowly as in March 1834, the Canal Company decided to carry out the work themselves. Bird was therefore given notice to quit, and he set up a boat building yard on the east bank of the canal below Llanthony Bridge. Unfortunately, the Canal Company then found that they could not go ahead with improving the graving dock as the Loan Commissioners would not authorise the expenditure. After struggling on for two more years, the dock gates were leaking so badly that the drain into the river had to be kept open permanently. This meant that when the river level was high, water flowed back into the dock and boat repair work had to stop. There were so many complaints that eventually the Loan Commissioners were persuaded to change their minds, and the much needed improvements were largely carried out in 1837. Surprisingly, no connection was made to the engine house at this time, and it could require up to eight men working for seven hours to pump the dock dry.

Meanwhile further warehouse building had continued, but with most of the space around the Main Basin now occupied on fixed term leases, it was increasingly difficult to find suitable sites. Thus in 1833, the timber merchant James Shipton built a warehouse beside the Barge Basin, and in the folowing year the corn merchants J. and C. Sturge built one by the lock. Neither of these warehouses had direct access to the Main Basin but this was a disadvantage which was accepted. To help relieve the pressure on space around the basins, the Canal Company made available a narrow strip of land near to Hempsted Bridge (the intervening land all being privately owned). The Reverend Samuel Lysons of Hempsted Court objected to this development as it was opposite an avenue of trees running down the hill from his house, and any building would spoil his view. He offered an alternative site on his own land slightly nearer Gloucester, but the Canal Company would not accept this as it was on the towpath side. They therefore went ahead with their original plan, and single storey warehouses were built for the Droitwich Salt Company and for the rival British Alkali Company circa 1836. Part of the agreement was that the canal would be widened and a quay wall built in front of each warehouse, and the earth removed was used as ballast for vessels leaving without any cargo. The Droitwich Company then sub-let their warehouse by the Main Basin, and their ability to do this profitably may well have been the motivation behind their move. During the same period, some of the merchants

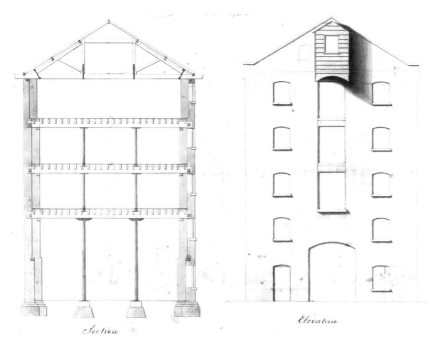

Section   *Elevation*

One of the original drawings for Shiptons Warehouse (dated 1833). The double headroom for the ground floor was unusual.

The salt warehouse at Hempsted Bridge built for the British Alkali Company of Stoke Prior.

pointed out the need for a shed on the public wharf to protect perishable goods during loading or shipment. The Canal Company first thought of erecting such a shed in front of their own warehouses, but Thomas Phillpotts and Samuel Baker came up with a better idea. They offered to provide a covered wharf by building a warehouse on pillars in the gap between the earlier warehouses on the West Quay and Sturge's new warehouse by the lock. In recognition of the public benefit, the Canal Company leased the land on favourable terms, and the warehouse was built in 1835.

There still remained a desperate shortage of waterside accommodation, and during the timber importing season, boats could wait their turn in the canal as far down as Hempsted Bridge.[7] When this happened, large pieces of wood were often lowered over the side and then chained together to form rafts that could be towed up to the timber yards by the basin when convenient. Sometimes balks were left in the water to prevent them drying out and cracking, but these tended to damage the banks and cause obstruction. Eventually, a special sub-committee was appointed to consider what improvements were needed, and they reported on various schemes in September 1835. These included two designs for a new dock to the east of the Main Basin and also a proposal to make a floating harbour in the river from Lower Parting to Westgate Bridge by building a lock at each end and by digging a new channel for the river across the meadows. After due deliberation, the proprietors concluded that the present state of the Company's finances could not support any of the schemes, and they suggested instead that it might be possible to let individuals form their own waterside accommodation. The urgency of the matter was accentuated when it was reported that there were over one thousand tons of timber floating in the basin and the canal, and the committee imposed a weekly charge to try to curb the practice.

Two specific proposals were soon put forward. A group of merchants and bankers led by Samuel Baker proposed to widen the canal and build a quay wall to the south of Llanthony Bridge, and Messrs. Price and Washbourne proposed to form a timber pond and deal yard further south. The timber pond would have relieved the congestion due to timber floating in the canal, but Price and Washbourne only had a short lease on the land they proposed to use and the scheme never materialised. Instead, they built a jetty on the canal bank to facilitate landing the timber. Baker and his friends, however, had bought the field known as High Orchard for their development, and they entered into an agreement with the Canal Company in which they promised to widen the canal by thirty feet and to construct a quay wall 360 yards long in

The Pillar Warehouse on Bakers Quay with Llanthony Bridge and the warehouses round the Main Basin behind (1843).

front of their premises. The agreement also stipulated that any warehouses were to be built on pillars to provide a covered wharf for the use of the public, and suitable roads were to be provided to link the wharf with Llanthony Road. There was also provision for a branch canal that could be built in the future to give access into High Orchard. Work on the new quay wall started at the northern end in 1836, and took several years to complete. It appears that the wall was built first and then the soil in front of it was removed with the help of a temporary tramroad which cut through the canal bank near Llanthony Bridge. Probably in 1838, the building now known as the Pillar Warehouse was constructed with the front of the upper storeys supported by pillars standing on the wall in accordance with the agreement with the Canal Company. The northern half of the warehouse was financed by Samuel Baker and the southern half by James Shipton, and both halves were used by the corn merchants J. and C. Sturge. The pillars proved to be convenient points to moor boats to, but following a complaint from Shipton, the harbour master was directed to stop the practice.

The southern end of the new quay was still not complete in 1839, and then the Birmingham and Gloucester Railway Company took up the option of forming a branch canal along the line of the Sudbrook. Their idea was to build a railway connection from their proposed terminus in Gloucester, and thus obtain direct access to the canal. The High Orchard Dock was completed in 1840 and sheds and coke ovens were built nearby, but the rail link was opposed by Worcester interests who feared a loss of their river trade and it was not authorised by Parliament.[8] Meanwhile, the remainder of the land behind Samuel Baker's new quay was mainly taken up by the timber trade, with James Shipton and John Forster occupying large yards, and with a saw mill and a timber preserving company setting up premises to the south of the branch dock.[9] The new facilities promoted additional traffic on the canal which could not otherwise have been accommodated, but the excavation in front of the quay had not been carried out properly. There were repeated complaints that the depth of water was less than that stipulated in the agreement with the Canal Company, so that large vessels could not get up to the quay and were instead obstructing the canal. This situation existed for over ten years until the Canal Company bought a steam dredger, and Samuel Baker and his associates paid for it to complete the work.

While the private developments were taking place to the south of Llanthony Bridge, William Clegram again pressed the Canal Company to develop their own land to the east of the Main Basin. In 1838, he pointed out the need to provide more wharfage for unloading coal going to Cheltenham on the tramroad and also for loading timber into barges that went on up the river. He recommended a new barge basin six hundred feet long and forty-five feet wide extending north from the existing Barge Basin. This would have given access to the rear of the timber yards along the east side of the Main Basin and hence reduced the existing congestion along the frontage of these yards. Once again, however, the Loan Commissioners would not approve the necessary capital expenditure and the idea had to be abandoned.

Opposite. Caustons map of 1843 showing the warehouses round the Main Basin and the yards served by the sidings of the Gloucester and Cheltenham tramroad. The field to the south of the shipbuilders yard was known as Berry Close and was the site of a later barge dock. To the south of Llanthony Bridge is Bakers Quay with the Pillar Warehouse and High Orchard Dock.

Another opportunity for development arose following the death of the tenant of the slate and bonded timber yards along the north side of Llanthony Road. In 1839, Clegram proposed widening the canal in front of the slate yard and building a small branch dock to give direct water access to the timber yard behind. Knowing that the capital expenditure would not be approved by the Loan Commissioners, a tenant was found who initially agreed to finance the development in return for a lease on favourable terms. Work started on the dock,[10] but the tenant objected to some conditions in the agreement and he backed out. The premises were advertised under the name Britannia Wharf,[11] but no one was prepared to finance the completion of the project and the excavation for the dock was eventually filled in again.

Far from being sympathetic about these missed opportunities, the Loan Commissioners became even more impatient about the money owed to them, and in 1839 they threatened to sell the canal unless definite arrangements were made to liquidate the debt. This forced the proprietors to review their finances, and they decided to raise sixty thousand pounds by the issue of preference shares. With Gloucester well established as a port for serving the Midlands and with toll revenue increasing, there was little difficulty in raising the money, and the Canal Company agreed to a regular set of payments for the remainder of the debt. This arrangement benefited the company by reducing the stranglehold that had been exercised by the Loan Commissioners, and it paved the way for the eventual distribution of a dividend.

CHAPTER FOUR

The Early Eighteen Forties

1840 was a particularly busy year, with little signs of the depression that was to follow. At times, the basin and the canal were crowded with vessels waiting their turn to unload, and the quays were bustling with activity. Large numbers of men were employed transferring sacks of corn into the warehouses, storing barrels of wine and spirits in the cellars and carrying long lengths of wood to be piled high in the timber yards. At the same time, other men were loading up smaller boats that would carry the goods further inland. The timber trade was particularly flourishing with an exceptionally high level of foreign imports which were needed for the construction of the early railways in the area. Much of the wood had to be cut up by hand, and as many as seventy pairs of sawyers could be seen at work in a field called Madleaze to the south of Bakers Quay.[1]

Corn imports were also increasing, and there was further demand for land around the basin for building warehouses. The Canal Company therefore arranged for James Shipton to give up his timber yard to the north of Biddles Warehouse, and they agreed to pay for moving his wood and his crane to Bakers Quay by way of compensation. Part of the space vacated was used in 1840 to build a huge new warehouse for Messrs. J. & C. Sturge. This firm had been remarkably successful since the opening of the canal, and judging from their warehousing capacity, they must have been responsible for about half the corn imported through Gloucester during this period. Just to the north, a warehouse was built at the same time for the Bristol based corn merchants Charles Vining and Sons. For the first few years, the building was shared with the iron merchants Britton and Darton, but later it was used exclusively for corn. Another firm of corn merchants from Bristol, Wait James and Company, were using one of the warehouses on the West Quay (both they and Vinings presumably being attracted to Gloucester by the better geographical position for supplying Birmingham).

In the following years, imports of both corn and timber were affected by the national economic depression, and some of the dock labourers and inland watermen were left without employment for months on end. The shipping industry was also depressed, and a new schooner built in William Hunt's yard to the south of the basin had to wait on the

Sturges Warehouse on the right and Vinings Warehouse on the left.

stocks a long time before a buyer was found.[2] This lull in activity relieved the pressure on the available facilities, and attention turned to the promotion of a larger export trade to reduce the number of visiting ships that had to go to other Bristol Channel ports to pick up return cargoes. The main existing export was salt from Droitwich and Stoke Prior, but this was more expensive than at Liverpool so that the volume of trade was not large. The timber merchants Price and Company managed to fill some of their ships on the outward journey to North America by carrying parties of emigrants who were trying to escape the hardships of the depression. Also, hopes were raised of a share in the export of manufactured goods from the Midlands when Henry Fox and Company sent a ship to Cuba with hardware from Birmingham and stoneware from Staffordshire, but this trade was dominated by Liverpool.[3] For a real expansion in exports, however, it was recognised that there was a need for a plentiful supply of coal, and the nearby Forest of Dean appeared to be the obvious source. A proposal was put forward for a railway linking Gloucester with the Forest, but this was complicated by interactions with other railway schemes and by the need to cross the River Severn without interfering with navigation, and it was to be many years before the line was eventually constructed.

Meanwhile, there were other moves to connect the docks with the newly opened terminus of the Birmingham and Gloucester Railway.

44

Having failed to get approval for a line to High Orchard, the Railway Company built a short length of tramroad in 1842 to connect their goods depot with the Gloucester and Cheltenham tramroad, and they rented the Britannia Wharf at the docks to facilitate the forwarding of goods to Bristol by boat. However, this route involved a laborious interchange between railway trucks and the narrow gauge tramroad wagons at the goods depot, and so in 1844 the Railway Company laid additional rails along the tramroad outside the tramplates to give a standard gauge line through to the docks.[4] New sidings and turntables were provided to serve the quays and the warehouses, but with traffic only being worked by horses and with some of the bends too tight for carrying long lengths of timber, this scheme was still unsatisfactory. In the following year, therefore, the Midland Railway Company (having absorbed the Birmingham and Gloucester) obtained parliamentary approval for a new line to the docks independent of the tramroad. It was particularly hoped that this would provide a means of increasing the export of salt, but there were further delays before anything was done.

One development that did go ahead during this period was the provision of improved customs facilities, as local merchants had complained that the existing office in town was inconveniently placed and was no longer adequate for the amount of business being transacted. A site was selected in what was to become Commercial Road,

The Custom House in Commercial Road.

and a fine new building designed by Sidney Smirke (brother of the better known Robert Smirke) was opened in 1845.[5] This was well placed to handle the huge expansion in foreign trade that was to take place in the following years.

Another forward looking move in 1845 was the opening of a steamer service to Chepstow by Henry Southan. Two other steamers were added later, and goods and passengers were carried between Gloucester and various South Wales ports. The new venture was welcomed by the Canal Company, but it took many months to evolve satisfactory methods of working. Unlike a sailing vessel, the steamers tried to operate to an advertised schedule, and this led to complaints about speeding on the canal, damaging a bridge and unloading at night and on a Sunday when the docks were supposed to be closed. The committee were particularly upset when a large number of people visited a new steamer one Sunday afternoon and porter was sold on board. After much correspondence between Southan and the Canal Company, it was agreed that two men would be employed on the steamers to regulate their speed and to help pass them through the bridges. Unloading at night was allowed and drinks could be served to passengers, but Sunday working was forbidden.

Vessels approaching Gloucester came under the authority of the harbour master, and it was his responsibility to allocate each one a berth or a mooring. To enter the basin, vessels passed through Llanthony Bridge, a wooden structure which was swung open in two halves by pushing on huge balance beams. There were no barriers to control access from the road, and children liked to ride on the bridge while it was being swung. One unfortunate pedestrian, whose attention was distracted by two men fighting, did not realise that the bridge was partly open and he fell off the end and was drowned.[6] Vessels were towed up the canal by horses, but all manœuvring in the basin was done by the crew using ropes and poles, and visitors were amazed to see huge ships gliding through the water without any apparent means of propulsion. Inevitably some mishaps occured, and contact with a moored vessel could result in an exchange of abuse between the crews and could occasionally lead to violence. One boatman used his boat-hook to push against the side of a French schooner, and a member of the schooner's crew threw a missile which hit him on the head. He was a big man, and he climbed on board the French vessel and chased those on deck with a rope-end until they all disappeared, some down the hatchways and others over the ship's side.[7]

Once a boat was properly tied up, there was an opportunity for the crew to go ashore, and most headed for one of the nearby taverns to

The original Llanthony Bridge, a wooden structure that could be swung open in two halves to allow vessels to enter the basin (c. 1840). The horse is standing on the towpath.

drink their fill and meet the local girls. Some then found that returning to their ship in the dark could be a problem, particularly if they had drunk too much. They might have to clamber across the lock gates, pick their way over the ropes criss-crossing the quays and clamber over other boats before reaching their own, and it is hardly surprising that some sailors ended up in the water. The problem was made worse by the lack of any proper lights, as there was a dispute about who was responsible for providing them. The Canal Company thought that gas lamps should be paid for out of the rates that were levied on the dockside property, but the local Commissioners would not install lamps on private property. There was just one oil lamp by the lock, but after various unfortunate deaths by drowning, the Canal Company installed other oil lamps by the graving dock and at Llanthony Bridge and later contributed towards the provision of gas lamps. Some sailors stayed ashore and only rejoined their ship as it was about to leave, but even this was not always without hazard. One sailor found that his ship was already going down the canal, and as he tried to climb aboard from a small boat, he fell into the water and got sucked under the ship and was drowned.[8]

Although most of the crew could go ashore, the Canal Company required that there should always be a competent person in charge of each vessel. If the cargo was dutiable, a customs officer also had to remain on board until the cargo was cleared. On one occasion, the original officer assigned to a vessel was relieved during the night by another officer who inadvertently woke the captain. Alarmed to find a stranger on board, the captain threatened him with two pistols, and the *Gloucester Journal* reported that 'the terrified revenue man beat a hasty retreat, glad to escape with such brains as he had in his head rather than have them bespattering the sides of the vessel.' The ejected officer had to maintain his watch ashore for the remainder of the night.[9] Lights and fires were supposedly prohibited on ships after nine in the evening, but breaches of this rule led to several incidents which could have been worse if they had not been spotted in good time. In one case, a fire was observed on a brig that had recently arrived with a cargo of sulphur and oil from Palermo. The fire was in the cabin and fortunately was extinguished before it reached the cargo, but a customs man on board was badly burned and later died in the Infirmary. Another fire occurred on a brig in the graving dock after the ship's boy had used a candle to light a lamp and had then gone off to visit another ship. Fortunately, the fire was spotted by one of the watchmen who sounded the alarm, and the city's fire engine arrived and extinguished the blaze before it did much damage.[10] The alarm bell used on these occasions was probably the one that hung from the corner of the North Warehouse and was normally used to signal the times for the dockers to start and finish work. The bell came from the ship *Atlas* which was launched at Whitby in 1812 but was not recorded after 1822.[11]

While vessels were in port, there was an opportunity for drying sails and for checking and repairing any damaged gear. Wooden sailing ships needed frequent attention to their hull and their rigging, and the services of the local shipwrights and sailmakers were often needed. There was particular activity in the spring when the locally owned timber importing ships were fitting out. (They were laid up during the winter because the Baltic and North American ports were closed by ice.) Major work on the hull usually required the use of the graving dock, but lesser matters could be attended to while a vessel was moored in the basin, and it was common to suspend a plank over the side of the boat to provide a working platform. One ship's boy caused quite a commotion when he untied a rope supporting such a plank, and three men on it were plunged into the water. Fortunately they were rescued safely, but the subsequent search for their tools brought up the body of a youth that had been in the basin for some weeks. It was presumed that the youth had belonged to a ship that had subsequently departed, and as desertion was not uncommon, no alarm had been raised at his disappearance. Repairs below the waterline could be carried out in some cases by rolling the vessel over on the earth bank to the south of the basin, but one schooner caused the whole bank to subside and two men working in the hold only just managed to get out before the boat sank.[12]

The bell which formerly hung on the east corner of the North Warehouse and gave the name to Bell Corner. It is now used at Shepperdine to guide ships in fog.

CHAPTER FIVE

A Major Expansion

By 1845, the economic recession had passed, the basins and the quays were crowded, the Company's finances were healthier than ever before and prospects for the future looked bright. The Anti-Corn-Law League was highlighting the harmful effect that the import duty on foreign corn had on food prices, and they were campaigning for its complete abolition. The government had already reduced the duty on timber, and so Gloucester merchants could look forward to an increasing demand for both of their main imports. With the imminent prospect of railway connections providing supplies of coal and salt, there was also every hope of expanding exports. The coastal trade was flourishing as well, with large quantities of iron, copper and tin being shipped in from South Wales, and the Severn Commission had been set up to improve the river above Gloucester which would help in distributing the imports. To cope with the expected growth in traffic, the Canal Company's engineer William Clegram again proposed the construction of a new dock to the east of the Main Basin, and there was pressure for a larger graving dock to accommodate the increased size of vessels now frequenting the port. The proprietors were initially cautious, however, and would only agree to spend money on buying a field called Berry Close to the south-west of the basin with a view to constructing a graving dock there later. The *Gloucester Journal* then started a campaign to promote the enlargement of the docks to take full advantage of being close to the consuming area of the Midlands and to compete with developments taking place at other ports. A series of editorials emphasised 'the absolute necessity which exists for arousing the minds of the commercial classes and the inhabitants generally of Gloucester from apathetic indifference at a moment when all the rest of the world are actively exerting themselves to improve their capabilities for carrying on trade'. The writer suggested that there was a need for new basins for transferring imports to railway trucks and for loading exports of coal and salt, and he suggested that the railway companies should be asked to help provide the capital. He particularly advocated the development of an export trade in salt to India, and he supported a proposal for a pipeline to bring brine from the Worcestershire saltworks to be made into salt at Gloucester.[1] This campaign evidently had some effect as, at

50

The basin crowded with vessels in the 1840s. The view is apparently from the field known as Berry Close, but some artists licence has been used.

their next meeting, the Canal Company proprietors asked the committee to look again at Clegram's proposal for a new dock.

In the summer of 1846, Clegram put forward three schemes. The first was for a large new ship dock to the east of the Main Basin and linked to it by a narrow cut. The second was for widening the canal opposite the Barge Basin and forming a quay with a barge dock running into Berry Close. The third scheme was to build a slipway for repairing ships on the east bank of the canal about two miles below Gloucester as this would be cheaper than a graving dock. At the next meeting of the proprietors in October, the committee expressed concern about 'the crowded and inconvenient manner in which the business of the docks has very frequently been carried on, the detention sustained by the shipping for want of berth room, the large quantities of grain that have occasionally gone to other ports that would have come to Gloucester had there been warehouse room for it and the several applications on the books of the Company for yards.' They also noted that Parliament had repealed the Corn Laws, and they expected that this would bring more trade to the port if adequate facilities were available. They recommended the plan for the new dock, but to save money they opted for a reduced scheme in Berry Close and proposed deferring action on the slipway. Samuel Baker suggested delaying any decision as the Gloucester and Dean Forest Railway were planning to build a huge new dock in Sizes Ground just to the south of Llanthony Road. The proprietors wisely felt, however, that they could not rely on railway company promises, and they endorsed the committee's recommendation. It was later agreed that the estimated cost of £16,000 would be raised by the issue of five per cent preference shares.

To remedy the shortage of warehouse space, work had already started on three new warehouses on the east side of the Main Basin. The southern one was built for the corn merchant Abraham Phillpotts, and when it was finished, he arranged for about seventy of the builders men to be 'bountifully entertained at a supper' at the White Swan Inn to celebrate the custom of 'house-rearing'.[2] The middle one was built as an investment by Humphrey Brown junior, and this was occupied by the corn merchant John Kimberley. Similarly, the northern one was financed by Samuel Herbert, a local solicitor, and this was occupied initially by Messrs. J. & C. Sturge. All three warehouses were built in 1846, and were ready to take advantage of the increased demand for foreign corn following the virtual abolition of the import duty. The land for the warehouses had formerly been used as timber yards by Price and Company, and they only agreed to move out on condition that they could expand their premises to the south of Bakers Quay by taking over the land used by John Bird for boat building. Bird was re-located to the south of Hempsted Bridge, where he constructed a small graving dock for the repair of trows and barges.

While plans for the main new dock were being finalised, work went ahead on the compromise proposal for bringing Berry Close into commercial use. The idea of a barge dock was retained but to save money no attempt was made to widen the canal or build a quay, and the sides of the dock were left as earth banks. Several wooden landing stages were built for boats to come alongside when unloading, and the whole scheme was completed for just over £1,000 by June 1847. This would have left the canal inconveniently narrow at the entrance to the new barge dock, but the tenant of the Britannia Yard opposite agreed to finance some widening on his side. In return for a twenty-one year lease, Thomas Tripp built a quay wall in front of his premises on the

Three warehouses built in 1846 to cater for the expected increase in imports following the repeal of the Corn Laws. (Herbert, Kimberley and Phillpotts).

same alignment as Bakers Quay, the work being carried out during the winter of 1847-8.

Meanwhile, Isaac Gaze had started work on preparing the ground for the main new dock. The first job was to clear a way round the east and north sides of the site for the Midland Railway's new dock branch, which was going to be built as an extension to their long delayed line to High Orchard. The Railway Company agreed to take away some of the excavated soil to help fill up some low ground on the south side of their station. Earth was also loaded into boats in the Barge Basin and taken down the canal to be dumped near the banks. As this work was proceeding, it revealed an unexpected difficulty in the form of a sewer running diagonally across the site from Southgate Street to enter the river by the lock. (This was probably the sewer that Dennis Edson had constructed after an earlier open ditch had caused flooding during the construction of the original basin.) It was agreed that an extensive culvert would have to be built in order to divert the sewer round the new dock. Before this could be done, however, Clegram reported that Gaze's work was not proceeding in a proper manner, and it was agreed that his contract should be terminated. The Canal Company took over his equipment on the site including miscellaneous planks, trestles, scaffold poles and ladders, twenty wheelbarrows, twelve ballast wagons, six hundred yards of iron rails and a three horse-power high pressure steam engine on wheels with a double acting pump and machinery for pile driving.

This equipment was then sold to the new contractor, William Guest, who started work in the spring of 1848. At this stage Clegram proposed some changes to the design of the dock to save expense. He reluctantly recommended abandoning the original proposal for a double swing road bridge across the entrance cut and making do with a foot bridge instead. With some of the money saved, he suggested widening the cut by fourteen feet except just at the entrance, thus providing additional quay space for small craft. Also, because of the culvert carrying the diverted sewer, the ground under the railway on the east side of the site had to be one foot higher than intended, and Clegram proposed to make the quay wall one foot higher as well. Although this would cause some inconvenience in the unloading of smaller boats, he thought it necessary in order to avoid an undue slope down to the quay. The construction of the culvert may also have been responsible for a change in the original rectangular plan of the new dock to allow the culvert to curve around the outside of the north-east corner. Work went ahead on the enormous excavation that was needed for the dock, and some of the earth was taken to Over Bridge for the railway embankment that was being formed there. This led to complaints that the carts being used were dropping soil in the streets of the town, and the drivers were warned by the magistrates against overloading and 'furious driving'.[3]

The construction of the dock probably followed the procedures laid down in the contract. This called for the excavation for the walls to be taken out in lengths to the full depth and properly shaped so that the brickwork could be built up solid to the undisturbed ground. The lower part of the walls were to be faced with Broseley bricks and the upper part with Forest or Bisley stone, and on the top a coping of Forest stone was to be secured with hard stone or slate dowells.[4] A simple ceremony was held on 30 May 1848 to mark the laying of the first stone, and a hogshead of brown stout was tapped for the consumption of the workmen.[5] When the construction of the dock was well advanced, a coffer dam was built in the Main Basin to allow work to proceed on the entrance cut. The contract specified that the dam should be formed by a double arch of twelve inch square fir piles with the space in between filled with puddle clay to make a waterproof barrier. Behind this protection, a section of the original quay wall was taken down and the cut into the new dock was formed. When all was ready, water was let in by sawing off the tops of two of the piles forming the dam. To complete the whole project, the Canal Company obtained two ten-ton manually operated cranes from John Stevenson of Preston.

The northern approach to the docks showing the entrance lock from the River Severn (c. 1850). On the left can be seen the lines of the Gloucester and Cheltenham tramroad and on the right in the distance is the chimney of the engine house.

While the dock was being constructed, progress was also made at last with bringing in a proper railway connection. The Midland Railway built a line from their station to their premises at High Orchard with a branch coming into the main docks area in 1848. This connected up wih the lines laid earlier by the Birmingham and Gloucester Company, and more sidings and turntables were added to serve the new dock.[6] With this access to the canal, the Railway Company had little further need for their dock at High Orchard, which apparently suffered from silting, and within a few years it was filled in and the ground was used as a goods yard. Traffic on the old horse-operated tramroad to Cheltenham was seriously affected by the arrival of the railway and by the loss of the northern end of the tramroad which was cut across by the excavation for the new dock. The sidings around the Barge Basin remained, however, and the line continued to give good service for several more years.

Another development at this time was the building of the Mariners Chapel. The Gloucester Journal estimated that the number of sailors and boatmen frequenting the docks could be over one thousand on occasions, and it noted that their life on shore caused 'much pain to every moral and religious mind.'[7] A few years earlier it had been decided to close the docks on Sundays to encourage the men to go to church, but few had taken the opportunity as they said that they did not have any smart clothes. The idea therefore arose of building a special chapel in the docks where sailors working clothes would not look out of place. Money was donated by the merchants and other public spirited citizens, a building designed by John Jacques was erected, and a special chaplain was appointed by the bishop to minister to the seamen and the dockland community. The opening service was held on the 11 February 1849, and subsequent services were well attended particularly on Sunday evenings. Unfortunately, it was found initially that a fair number of people from the city were also coming, and they had to be encouraged to return to their own churches for fear that their presence would deter the seamen.[8]

The formal opening of the new dock took place on 18 April 1849 and was watched by thousands of spectators. When the proprietors arrived, they were greeted by the bells of St. Mary de Crypt and the firing of cannon, and Union Jacks flew from the tops of the warehouses. Ten vessels had assembled in the Main Basin with long lines of flags in their rigging and with national flags at their mastheads. The first boat into the new dock was the brig *Torrance* of Wexford laden with corn from Constantinople, and she made a fine sight with her crew manning the yards and with two men at the mastheads. As she moved into the dock, the seamen and spectators cheered, a band played Rule Britannia and the cannon 'roared a salutation of smoke and fire.' She was followed by the other boats, and as they reached their berths, some were boarded by spectators while others started unloading. There was competition to claim putting the first parcel of goods ashore or the first bag of grain, and a hearty cheer was raised when the first railway truck was filled with a load of paper. To complete the day, a sheep was roasted and several barrels of beer were distributed amongst the contractors workmen. The *Gloucester Journal* reported thankfully that in spite of the large crowd, nobody met with the slightest accident.[9]

The new dock and the land around it were soon being put to good use as Gloucester benefited from the government's Free Trade policy and from the repeal of the Corn Laws in particular. The Victoria Warehouse was built in 1849 for William Partridge, a carrier between Birmingham and Gloucester who also traded as an iron merchant. He leased the warehouse to Wait James and Company, a long established firm of corn merchants who had previously occupied a warehouse on the West Quay. Two years later, William Partridge also financed the building of the Albert Warehouse, and this was leased to the corn merchants W.C. Lucy and Company. Meanwhile, Joseph and Jonah Hadley had built the City Flour Mills in 1850 just to the north of the new dock. The plant

The Mariners Chapel in the 1880s.

initially consisted of a few pairs of stones and some flour dressing
machines driven by a steam engine. This was an early example of the
gradual movement of the corn milling industry from water-powered
sites in the country to steam-powered mills where there were good
transport facilities. The venture was so successful, that the Hadley
brothers soon built an adjoining warehouse, and they installed two new
steam engines and other machinery to double the output of the mill. To
the south of the Victoria Warehouse, the Hull based firm of Barkworth
and Spaldin formed a bonded timber yard, but the Canal Company
insisted that the lease could be terminated at six months notice so that
they could make the land available for further warehouse building if
needed. Across on the east side of the dock, the Midland Railway
Company laid a special siding to facilitate the shipment of salt, and the
Canal Company agreed that the adjoining berth should generally be
reserved for this purpose. Nearby, Jesse Sessions established a coal and
builders merchants yard on what became known as the Albion Wharf
(opposite the Albion Hotel in Southgate Street.) The new dock was

known both as the Victoria Dock and as the New Basin, and the Canal Company also referred to it as the Southgate Street Dock for several years.

The proprietors of the Canal Company were delighted by the successful completion of the new dock, and were undoubtedly pleased to see the provision of more warehouses and timber yards and proper railway facilities. They recognised the important roles that their engineer and their clerk had played in designing and supervising the construction of the dock, and both men were thanked for their able and long continued services to the Company. The proprietors marked their gratitude by presenting William Clegram senior with a silver tea service, and by giving his son a pair of silver candlesticks and a microscope.[10]

One disappointment was that the Gloucester and Dean Forest Railway Company had made only slow progress with their line to the west of Gloucester, and so there was still no export trade in coal to balance the growing imports of corn and timber. It became clear that the Railway Company would not have enough money left to build their planned dock to the south of Llanthony Bridge, and it was agreed instead that they would just widen the canal and construct a quay wall 360 yards long opposite to Bakers Quay. This required the demolition of the original bridge-keeper's house and so a new house was built on the east side of the bridge. The quay wall was constructed in a huge trench which was dug parallel to the canal, and the original canal bank was left in place to keep the water out. Unfortunately, the bank was weakened by heavy rain and the wash of a steamer started a leak. Frantic attempts to stop the leak were not successful, and soon the whole trench was flooded and the workmen's platforms were 'floating in profusion in the seething waters'. One man had a narrow escape as he jumped down to recover his pick and only just managed to clamber out in time as the water rushed in. This was a serious set-back for the contractor, but he repaired the breach and pumped out the water, and soon had his men back building the wall again. When the wall was finished, the water was let in deliberately, and then the earth bank was removed. Initialy this was done by men who had to work up to their waists in water, but once they had gone deep enough, the Canal Company's newly acquired steam dredger was brought in to finish the job.[11] The quay wall was completed in 1852, but there was still further delay before the railway connection came into operation, and in the meantime Llanthony Bridge had to carry much heavy traffic to and from the new quay. This put such a strain on the old wooden structure that a weight limit of four tons was imposed, and eventually the bridge had to be closed temporarily for repairs.

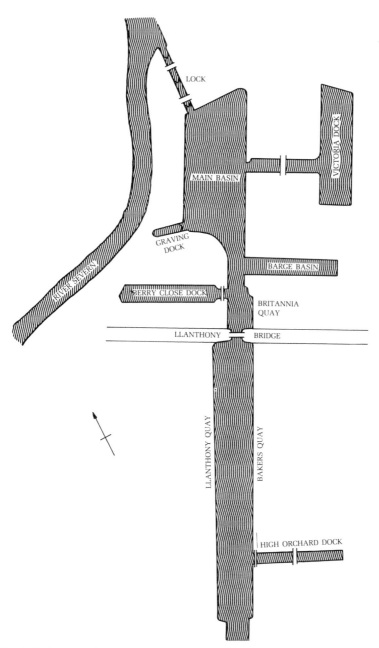

The docks in 1852, showing the expansion that had taken place in the previous five years by the building of Berry Close Dock, Britannia Quay, Victoria Dock and Llanthony Quay.

CHAPTER SIX

Fluctuating Fortunes

With the benefit of the improved facilities at Gloucester, trade increased steadily during the early 1850s, and on several occasions nearly reached the limit of the canal's capability. Corn imports were so high that it was sometimes difficult to find sufficient storage space even with the recently built warehouses by the Victoria Dock. In August 1853, there were thirty vessels lying two or three abreast below Llanthony Bridge waiting to enter the basin. Also timber imports were exceptionally high due to the railways being constructed in the Midlands, and two railway contractors, William Eassie and Tredwell and Company, set up yards between the Bristol Road and the canal for cutting and shaping the timber they needed. To serve Tredwell's new premises, the Canal Company tried in 1853 to arrange for an extension of the railway down the east bank of the canal, but Price and Company initially refused to allow the line to cross the yard they rented south of Bakers Quay. The Canal Company ordered their solicitor to start legal proceedings, and after six months of wrangling, a compromise was reached. Price and Company agreed to withdraw their opposition on the condition that the Canal Company would excavate lie-byes in the canal bank suitable for taking two large vessels an also that they would install piling along the canal bank. This opened the way to the construction of a new quay 900 yards long in front of Price's and Tredwell's yards, and the railway laid along it was connected with the Midland Railway at High Orchard.

To improve communications with Sharpness, a number of new passenger steamers started operating on the canal during the early 1850s. After a special trial to check on possible damage to the banks, the Canal Company licenced the *Express* and the *Wave* to run regular services, and they licenced the *Spitfire* to meet the tides at Sharpness. A fourth boat was refused a licence because it made too much wash. Regulations were laid down to control the times of departure and the speed of the boats, and they were prohibited from sounding their whistles or blowing off steam between Frampton and Purton as Earl Fitzharding had complained that it upset the game around his decoy pool. Great rivalry developed between the different boats, and the *Express* and the *Wave* were caught racing on one occasion. There was so much squabbling in fact, that in 1853 the Canal Company came to an exclusive arrangement with one operator, J.G. Francillon, the owner of the *Wave*. He then acquired the *Lapwing*, and these two boats started a regular service between Gloucester and Sharpness that was to last for many years. The passengers included merchants visiting Sharpness to inspect their cargoes as they arrived, sailors joining their ships at the last moment before departure, and a growing number of labourers who worked at Sharpness and only returned to Gloucester at the weekends. The little packet boats also provided transport into Gloucester for the inhabitants of the rather isolated villages along the line of the canal, and during the summer they were a means of giving several thousand city dwellers a pleasant day out in the country.

Although imports were booming, the number of outward cargoes remained disappointingly low. Salt was still the only significant foreign export, but special arrangements were also made occasionally. For example, Price and Company continued to use their timber ships to carry emigrants across the Atlantic. Most of the parties were made up of families with their children, but on one trip the 250 passengers were mainly young paupers from Cheltenham who had been brought to Gloucester in a special train. The chaplain of the Mariners Chapel always went on board prior to departure, and he spent some time talking with individuals and distributing tracts before holding a service in which all the emigrants usually joined.[1] In 1853, Humphrey Brown started a line of ships carrying goods and emigrants to Australia following the discovery of gold there. He bought ships from America, Russia and Nova Scotia as well as two that were built in Cardiff, and the first carried watches, boots and shoes, wine, beer and pre-fabricated wooden houses, all from local traders.[2] Brown aimed to send out one ship a month, but the scheme was not a success, and he soon had to sell up and later went to prison for embezzling £77,000 from the bank of which he was a director. In the coastal trade, there were some outward shipments of home grown oak for the naval dockyards, and a special timber yard was established near Hempsted Bridge to store the oak before dispatch. For a substantial export trade, however, what was still needed was a cheap supply of coal, and so there was much interest in the progress of the Gloucester and Dean Forest Railway line that was at last being constructed to serve the quay that had been built just below Llanthony Bridge.

With a real prospect of a railway connection giving an export trade in coal, the Canal Company turned their attention to the requirement for a new graving dock. The larger vessels coming to Gloucester were having to get repairs done at some other port where they went to pick up a return cargo, and this was not compatible with the hopes for a growth in export trade. Contacts were made with shipwrights in Sunderland and Newport who might be prepared to finance the dock, but they were only interested if they could have a long lease that would give them a virtual monopoly of the ship repairing business of the port. The Canal Company therefore decided to build their own dock to be an amenity that could be rented by anyone who had need of it. Clegram drew up a design suitable for the largest ships that could come up the canal, and a site was selected in Berry Close adjacent to the existing graving dock. It appears that the nearby Berry Close dock was filled in to make space for the new dock, although no specific reference to this has been found. The contractor was William Guest, and the Neath Valley Iron Company supplied a steam engine to pump the dock dry when the river was too high to allow the water to drain out naturally. As the work was nearing completion, the landlord of the nearby Black Swan was given permission to hold a party in the dock but there is no record of whether such an extraordinary event ever took place. The dock was first used on 4 August 1853 when a 350 ton barque was hauled in to be stripped, caulked and coppered. The *Gloucester Journal* reported that the occasion was celebrated by flags floating from the rigging and by the firing of cannon, and the captain supplied his crew with port wine and the main local shipwright gave his workmen a good supper.[3]

The long delayed branch of the Gloucester and Dean Forest Railway serving the quay below Llanthony Bridge was eventually opened in 1854. To assist in transhipping the coal, a powerful hydraulc lift was constructed by the Great Western Railway who had taken over operation of the line. Each railway truck was placed on a platform which was raised by four strong chains, and then the platform was tilted to let the coal tip into the hold of a waiting vessel. The lift was capable of raising thirty-six tons, and the power was provided by a steam engine at some distance from the quayside. This was an early example of a hydraulic lift and it was said to be the most powerful machinery of its kind so far introduced. The lift was first used to load 120 tons of coal from Messrs. Nicholson's colliery at Parkend on to an Irish schooner bound for Wexford. The process was evidently rather messy, as the *Gloucestershire Chronicle* warned those promenading along the canal banks 'that the operation will in no way add to the pleasure of their walk'.[4] In practice, however, the walkers were not disturbed much as the railways opened up a wide home market for Forest coal, and the demand was so high that little was available for export. The railways were also responsible for reducing the amount of coastal trade, which by the 1850s was largely confined to interchange with other ports in the

The large graving dock in 1936. On the right is the steam chest used for bending timber.

Bristol Channel, and some people began to wonder whether railways could be more of a threat than a benefit to the prosperity of the port.

The large number of ships coming to the docks brought quite an influx of visiting seamen. Most stayed only a few days and never returned. Others, such as the Irish, were regular visitors and they often stayed for three or four weeks at a time. The French sailors were viewed with suspicion as they were seen going out in the early morning with sacks, and it was said that they were gathering snails to eat. Many of the seamen spent their money drinking and gambling in the taverns near the docks, and they were often involved in fights with the locals. They were sometimes robbed of their wages with little chance of justice as they usually had to sail before any accused could be brought to trial. The chaplain of the Mariners Chapel spent much of his time trying to combat these problems. He made a point of visiting each ship when it arrived, and he often called at the two sailors homes in the town to spread the word of God. He sold bibles, distributed tracts in many different languages and encouraged the men to give up drink and gambling and to come to the chapel. Sunday services were usually well attended, and special services in foreign languages were arranged occasionally. For those who were still unwilling to come into the chapel, the chaplain sometimes took out a portable organ and held a service by the dockside or on board a ship. It was often frustrating work as few of the sailors stayed in port for long, but it was clearly appreciated by many. The nucleus of his congregation was drawn from the dock labourers and their families, and for them he organised bible classes and an adult evening school. A highlight of the year was the annual tea party which was held in one of the warehouses and was attended by several hundred people. One group of workers on whom the chaplain had little influence was the inland waterways boatmen. Their demoralised condition was proverbial and the chaplain attributed this to them still having to work on Sundays.[5]

The prosperity of the early 1850s was interrupted by the outbreak of the Crimean War in 1854. This particularly affected Gloucester as most of the corn imports came from the Black Sea area and much of the timber imports came from north Russia, and these were stopped by the war. There was high unemployment among the dock workers, and many went to seek work elsewhere leaving their families to go into the workhouse. Joseph Sturge kept some of his men employed cleaning and whitewashing the empty warehouses, but this attitude was exceptional. The problem was made worse by the hardest winter for thirty years which froze up the canal for over two weeks in February 1855. During this period, a sheep roast was organised at the docks, using two ship's stoves mounted on some thick planks on the ice. Over three thousand people gathered to watch, and the ice became so crowded that the water began to lap up over it and the sheep had to be moved hurriedly to a timber yard just to the south of the Barge Basin. The Gloucester Journal reported that many hundreds of people had a taste of the meat and more would have liked some even though it was not fully cooked through.[6]

In spite of the difficult times, the Canal Company realised that they had to do something about improving the water supply to the canal. The original pumping engine had been modified to help cope with the pre-war increase in traffic, but in the exceptionally dry summer of 1854, it was not able to prevent the water level in the canal dropping two feet below normal, and large ships had to transfer part of their cargo to lighters at Sharpness before proceeding up to Gloucester. While the Canal Company were considering what to do, they were approached by the local Board of Health who were in even greater difficulties. The city's reservoirs on Robins Wood Hill had run dry, and the Board of Health asked if the pumping engine could provide an emergency supply from the River Severn. Although there was considerable concern about the purity of the river water, particularly as the pump inlet

An artists impression of the docks in 1867 although the exact location is not identifiable. The nearest vessel has a temporary platform under the bow to facilitate unloading timber through a special port.

was downstream of a sewer outfall, two serious fires in the town had convinced the Board that they had to act. The Canal Company agreed to help for a limited period, and an additional pump was attached to the engine, pipes were laid crossing the canal at Llanthony Bridge, and in just over a month the first water was pumped up to Robins Wood Hill. There were some difficulties due to burst pipes and due to unauthorised operation of control valves, but the scheme provided much needed relief to the city for several months.[7] Meanwhile the Canal Company was arranging to obtain a new engine from the Neath Abbey Iron Company and to modify the engine house to take it. The engine had a beam weighing ten tons and a flywheel that was eighteen feet in diameter, and two pumps could supply 10,000 gallons of water per minute. All was ready by November 1855, but then there was a set-back when the cast-iron flywheel shaft broke in two during a trial. Clegram recommended replacing it with a wrought-iron shaft, but the Neath Abbey Iron Company blamed the trouble on the foundations, and the brick supports for the shaft were eventually replaced by stone.

After the war, trade gradually recovered, and there were hopeful signs of new opportunities in 1856, when a ship arrived with a mixed cargo from India and another in the same week brought bones from South America. In the following year, a schooner with an auxiliary steam engine arrived with barley from Stettin, and it steamed up the canal under its own power.[8] These developments were encouraging, but it was also becoming apparent that the railways were eroding Gloucester's geographical advantage of being the most inland port. The Midlands could now be supplied from London, Liverpool or Hull, and these ports had provided facilities for the larger sizes of sailing ships and steamers coming into use. Consideration was given to enlarging the canal entrance at Sharpness, but the cost estimates were very high and there seemed little chance of an adequate return on the investment. There were suggestions that public money should be made available as the recent war had brought home to people the importance of the canal and the docks to the prosperity of the city, but the whole matter was deferred as Gloucester began to share in a general growth in the nation's overseas trade.

CHAPTER SEVEN

'Nowhere is Any Inertness Visible'

By 1860 trade was flourishing again, and steam tugs were brought in to replace horses for towing along the canal. One tug could tow two or three large vessels or as many as ten smaller vessels, and this gave a considerable cost saving compared with using horses. The tugs were also a great help in coping with the increased volume of traffic. In one week in November, more than eighty vessels entered the port. Half of these brought corn from Ireland or the continent, four brought corn from the Black Sea area and one from New York. Eight brought timber and deals from the Baltic and one from Canada, and there was wine from Opporto, sulphur ore from Pomaron, glass sand from Rouen and sundries from Holland. In the coastal trade, vessels brought coal from South Wales, slates from North Wales, pig iron from Scotland and porter from Ireland.[1] The *Gloucester Journal* reported with pleasure that the basin

'is now so crowded with vessels of all descriptions, that some difficulty is experienced by captains and owners in safely removing their charges from the labyrinth of small craft, spars, cordage etc. with which they are speedily hemmed in. There is a continuous babel of strange voices and discordant cries ... while shrill answers are returned to hoarse demands from mast heads, cross trees, dark cabins, holds, cabooses and stern ports. Nowhere is any inertness visible ... Cranes yell shrill remonstrances at the heavy sacks of grain which they are perpetually removing from gloomy holds, cordage creaks with the weight of pine brought from the impenetrable forests of Norway, while dingy, bearded foreigners utter strange ejaculations as they stagger under the products of their native climes. Here, we see a Frenchman from the rich vine districts of Brittany, or an Italian from the fertile plantations around Palermo. Again, a swarthy Negro escaped from the Slave States of America, and an occasional Malay. These, with a few Americans and a sprinkling of Norwegians, Danes, Dutchmen and Germans, compose the motley crews of the arrivals in our port. It is refreshing to find that the numerous "ne'er do wells" who previously idled round the basin have found occupation in transhipping grain and timber, while those workmen to whom labour is the vital element of existence manage to make sufficient overtime to gladden the heart of many a thrifty housewife, and garnish the domestic board with luxuries as unusual as they are welcome.'[2]

With this mixture of nationalities, it was not surprising that some difficulties occured occasionally. When a gang of corn porters went on board an Austrian barque to help with the unloading, they found that the crew were mixing some damaged grain in with the good grain instead of keeping it separate as was the usual custom. The porters objected, but the two sides could not speak each others language and a fight broke out. The Austrians rushed forward, but they were quickly knocked to the ground by the superiority of the English punching. According to the *Gloucester Journal*, arms fists and legs flew about in mad confusion, and the shouting and swearing drew a large crowd from all parts of the docks. The fighting eventually subsided as the Austrians offered no further resistance, and two of them were subsequently taken to the Infirmary for treatment.[3]

The high level of activity in the docks must have been particularly gratifying to William Clegram, who had been the Canal Company's engineer and general superintendent for over thirty years. He had started out as a sailor, but had changed to engineering when he won the competition for improving the harbour entrance at Shoreham. Since coming to Gloucester, he had probably done more than anyone else to promote the development of the docks, and he was admired by the merchants for the attention he gave to everything that would facilitate their trade.[4] He was well over seventy, however, and ill health eventually forced him to resign his duties in 1862. He was succeeded by his son, William Brown Clegram, who had worked with his father on engineering matters while serving as clerk to the Canal Company.

With the return of confidence after the war, investment in buildings started again. The Britannia Warehouse was built beside the Victoria Dock in 1862, just to the north of the link with the Main Basin. It was the third to be financed by William Partridge and it was leased to the corn merchants Henry Adams and Company. In the following year, Messrs. T.N. and R.G. Foster moved from Evesham and built an oil and cake mill at the southern end of Bakers Quay, where imported seeds could be unloaded direct from sea-going ships. On the front of the building was a fine looking timber structure which projected forward and was supported on pillars in accordance with the original agreement concerning the construction of Bakers Quay. Inside the building, large hydraulic presses were used to crush linseed and cottonseed to

Foster Brothers Oil and Cake Mill designed by George Hunt of Evesham. The wooden structure on the front supported by pillars had to be taken down when the quay wall collapsed in 1892.

give oil, and the wide flat slabs of residue were sold for cattle food.

The shipbuilding yard beside the large graving dock in Berry Close was in action again after being closed for some time. The Canal Company had arranged for the Sunderland firm of Pickersgill and Miller to take it over in the belief that they would also help to provide the skills needed for an efficient ship-repairing service. Pickersgill and Miller had some difficulties in launching some of their boats, and this was attributed to the boats being fully rigged before launching as was the practice in the north of England. In July 1861, a barque called *Mary Stow* stuck on the slipway and a steam tug was brought in to help. A large crowd of onlookers had gathered, and in the excitement a mother dropped her baby into the water. Fortunately the baby was soon rescued, but the occasion proved to be too much for the harbour master who had a heart attack and died. The barque was eventually launched on the following day with the steam tug helping again, but those on board had a bad fright when the boat lurched to one side and almost capsized as it entered the water.[5] Pickersgill and Miller also established a shipbuilding yard to the south of Hempsted Bridge near to the small graving dock formerly used by John Bird. They built three-masted barques and schooners of two to three hundred tons burthen, and these were sold to Bristol and Liverpool owners for trade to China and South America etc. The firm also built at least one steam yacht, but this got them into trouble with the Canal Company who complained that it had run at excessive speed on the canal.

The recovery in trade also prompted another set of improvements to provide more quay space and warehouse land. In 1861 the short stretch of canal between the Main Basin and Llanthony Bridge was widened by Joseph Fowler of Ross, a quay wall was built in front of Berry Close and work started on extending the Britannia Quay northwards to the Barge Basin. Work also started on replacing the old wooden Llanthony Bridge with an iron swing bridge so that the Midland Railway could lay a line across it and into Berry Close. This was delayed, however, when it was found that the east pier of the old bridge was defective, and a coffer dam had to be constructed to allow a new wall to be built round the pier. While the bridge was out of action, a rope-operated ferry was arranged for pedestrians, but its handrails and steps for boarding did not appear very secure and it was 'entirely avoided by females'.[6] All the improvements were eventually completed in 1862, and the new bridge was tested with a load of twenty-six tons. The Great Western Railway also laid lines into Berry Close and along the road behind the warehouses on the West Quay, and the Canal Company built a wall to protect the bank of the River Severn. With these improvements, the land in Berry Close became the site for another warehouse and further timber yards. The warehouse was built in 1863 just to the north of Llanthony Road and was the fourth to be financed by William Partridge. It became known as the Great Western Warehouse and was leased to the corn merchants W.C. Lucy and Company. It was a particularly large warehouse and was expected to provide valuable protection from south-westerly gales of the kind that had caused a brig to capsize in the basin many years earlier. Also in 1863, the south wall of the barge basin was raised by two feet to allow the Midland Railway to lay sidings to handle the road-stone traffic that had formerly been carried on the Gloucester and Cheltenham tramroad. The coal traffic on the tramroad had fallen away in the face of competition from the railways, and complaints about the bad state of the rails running unguarded through the streets had led to the line being abandoned.

An improvement of a different kind was the lighting of Bakers Quay by gas. With easy public access, deaths by drowning were fairly common in this area, and the local coroner had been pressing the Canal Company to do something for several years. There were cases of suicide and of children being drowned while playing on moored boats, but the main problem was sailors returning to their ships at night while under the influence of drink. The Canal Company did not feel that they were responsible as Bakers Quay was not their property, but under pressure from the coroner they encouraged the timber merchants who had yards there to get the lighting installed. They also provided each bridgeman with a lifebuoy and procured more drags for recovering bodies. When in 1863, therefore, the captain of a smack was drowned while returning to his boat at Bakers Quay, the Canal Company were dismayed to find that they were again criticised for not providing lights or restricting access to the quay at night. They quickly wrote to the *Gloucester Journal* pointing out that lights had been provided and that the captain had every right to be on the quay.[7] After this, it seems to have been accepted that some accidents were inevitable considering the large number of visiting seamen and their partiality for alcohol.

Alcohol was also the reason for some sailors ending up in court. For minor offences, the charges were not always proceeded with, as the magistrates were prepared to accept assurances from the accused or from his captain that he would be leaving port soon. More serious cases were committed to the assizes in the usual way. In 1863, a German sailor was given three months imprisonment for stealing gold and silver watches and jewellery from the Sailors Home where he was staying. Having sold the articles in town, he had caught a train for London, but the police telegraphed to Swindon and he was arrested there. Pilfering at the docks was another matter to come before the courts, although it was sometimes difficult to prove a case due to vital witnesses having to leave before the trial. Two corn porters were arrested after the cargo of American wheat that they were unloading was found to be twenty quarters short and they were seen taking wheat to two local mills. At the assizes, however, they said that the captain had given them the wheat, and as the captain had long since returned to sea, the judge ordered the men to be acquitted.[8]

An inconvenience that was becoming increasingly troublesome, both to local workers and to the crews of visiting ships, was the lack of a proper supply of drinking water in the main docks area. It seems that some crews made use of taps on nearby private property, but this was quite unofficial and not without risk. The *Gloucester Journal* reported that one Irish sailor had filled a cask from a forbidden tap, and as he was hurrying back to his ship with the cask balanced on his head, he had a nasty shock. The bottom of the cask suddenly gave way and the contents deluged all over him, much to the amusement of several onlookers.[9] The merchants also wanted a supply of water for the benefit of their workers, and in 1863 they asked the local Board of Health to lay down water mains, pointing out that the property around the docks already contributed largely to the rates. In response to this request, the Board installed a public drinking fountain near to the Canal Company's office. It was soon found, however, that the fountain was so frequently used for filling casks for ships going on long journies that the dock workers for whom it was intended could not get proper access to it.[10] To meet this difficulty, a second drinking fountain was eventually provided near Llanthony Bridge for the benefit of the visiting sailors.

During the 1860s, the tonnage carried on the canal was much the same as it had been in the 1840s, although the total now included a far larger contribution from foreign imports and a reduced contribution from coastal and inland trade which had been affected by railway competition. The corn trade was particularly healthy and had expanded three-fold since the import duty had been cut by the repeal of the Corn Laws. As well as a flourishing trade with Ireland, there were many vessels bringing cargoes from the Black Sea area, France and other north European countries. In one year, over three hundred French vessels came to the port, and the *Gloucester Journal* noted that groups of Frenchmen could be seen in town clattering their wooden shoes and jabbering and gesticulating to the great amusement of simple country folk.[11] The old established firms of J. and C. Sturge, Wait James and Company, Phillpotts and Company, J.P. Kimberley and C.J. Vining had been joined by new firms such as W.C. Lucy and Company, J. and T. Robinson, John Weston and Company and Henry Adams and Company. Each firm had its own warehouse, and they also rented additional space in the Canal Company's North Warehouse when needed. The throughput of corn was the equivalent of that required to fill and empty each of the warehouses three or four times a year. At peak periods, merchants could have great difficulty finding sufficient storage, and some consignments had to be turned away to other ports. Many shipments of French barley were transferred directly to the Midland Railway to be sent on to Bass and Company of Burton for conversion into malt. Unfortunately, the transfer was often delayed by a shortage of railway wagons, and when some of the barley deteriorated by being left

The drinking fountain provided by the local Board of Health.

in the open, the traffic was switched to Newport. This caused consider-able consternation, and in 1866 the Canal Company built an iron-frame transit shed to the east of the Victoria Dock for the 'special accommoda-tion of the grain trade'. The Midland Railway built a similar shed in their yard at High Orchard.

The timber trade was also expanding, although imports did not recover to pre-war levels until the late 1860s. Most of the timber came from the Baltic or Canada, but there were also consignments from the United States, Norway and the Arctic coast of Russia. The principal merchants were Price and Company, William Nicks and Company and Barkworth and Spaldin, and the yards were concentrated along Bakers Quay and the canal bank to the south. Other foreign imports included wines and spirits (particularly brandy), oranges and lemons, bones and guano for fertiliser, rags for paper and occasional consignments of wool. Salt was still the only regular export, and most vessels had to go elsewhere to find a return cargo, particularly to the Welsh ports to pick up coal. This imbalance of trade was said to be the reason for the shortage of railway trucks needed to distribute imports, but all attempts to encourage more exports proved fruitless. The main cargoes in the remaining coastal trade were slates, coal and iron, with some salt going outwards.

In the early days, most of the imports brought up the canal were sent on to be processed by small firms serving their own locality in the industrial towns of the Midlands. With the growth in trade as the century progressed, however, there was a tendency for processing industries to be set up near the docks to take advantage of the direct transport facilities. The City Flour Mills and Foster Brothers oil mill have already been noted, and there were timber sawing and planing mills in the High Orchard area operated by Price and Company, William Eassie and Company and Samuel Moreland. Moreland later specialised in making matches, and he established a factory in Bristol Road not far from the canal. The Gloucester Wagon Company also had their works in Bristol Road, and they later took over Eassie's premises

The Albert Flour Mills (c. 1900).

St. Owens Mills by the small graving dock (c. 1900).

and established their own water frontage just to the south of Bakers Quay. During the 1860s, several more steam powered flour mills were started in Gloucester and two of these were in the docks area. About 1863, William Hall and Sons established St. Owens Mills in a former warehouse on the West Quay of the Main Basin, and in 1869 James Reynolds converted the Albert Warehouse by the Victoria Dock into a flour mill. The good transport facilities available to these dockside mills was a distinct advantage, and helped to keep them operating long after most of the inland mills had closed.

The lifeboat behind the North Warehouse being prepared for the procession through the town.

A special event took place on 9 April 1867 when a new lifeboat was officially launched and handed over to the National Lifeboat Institution. The money for the boat had been raised locally, and it was to be named *The Gloucester*. After a civic luncheon, the boat was drawn through the specially decorated streets in a long procession which included the mayor, four bands, three fire brigades, four units of the militia and several other groups. Only ticket holders were supposed to enter the docks area, but the gates were forced open and a crowd of over 8,000 swarmed in round the Victoria Dock. The *Gloucester Journal*

The ceremonial launching of the lifeboat. The spectators made use of every available vantage point, however perilous.

reported that people made use of every available vantage point however perilous, and some brought piles of wood from a nearby timber yard to help get a better view. After a few formal speeches, the boat was launched with the crew on board, thereby creating an enormous sheet of spray. It was rowed round the dock rescuing volunteers who had thrown themselves into the water, and it was then capsized by a crane to demonstrate its self-righting ability. Afterwards, the streets remained crowded and the resources of the inns and refreshment shops were said to have been seriously taxed.[12]

The canal crowded with vessels in 1867. The brig Ada of Shoreham had brought 370 tons of linseed from Taganrog on the Black Sea for Foster Brothers mill. Standing by the longboat is a member of the Butt family who became coal and builders merchants in Stroud.

CHAPTER EIGHT

The Challenge of Competition

The high level of activity in the docks throughout the 1860s allowed the Canal Company to distribute a modest dividend to shareholders as well as to pay off at last most of the money originally borrowed to finance the completion of the canal. It became apparent, however, that a major new development would be required at Sharpness, as the size of ships was increasing and many were too big to come up the canal fully laden. These had to transfer some or all of their cargo to lighters in the tidal basin at Sharpness and this practice caused considerable congestion. In 1869, the Company's engineer, W.B. Clegram, reported that ships were being delayed between twelve and twenty-three days waiting their turn to be admitted to the basin, and the detention of outward bound vessels was nearly as great. He also pointed out the need to provide facilities for steamers which were increasingly being used for the conveyance of grain to England, or else the foreign trade of the port would inevitably suffer.[1] He suggested a new entrance, tidal basin and dock at Sharpness that would take the largest ships of the day, and after much deliberation the proprietors decided to go ahead.

Work on the new dock started in 1871 but not without misgivings in some quarters. The *Gloucester Journal* published a prophetic dream which predicted that Gloucester would be defunct as a port by the turn of the century and that all activity would have transferred to Sharpness. On an imaginary visit in the future, the dreamer found that the docks at Gloucester

> . . . were deserted, and contained no water, with the exception of here and there a stagnant puddle covered with verdant duck-weed, in which gamboled the tadpole and stickleback. The quay walls had fallen in, the bottom was covered with old boots, shoes, crocks, pans, rusty pots, kettles and fragments of multitudinous utensils of a departed generation. The warehouses here and there had become huge unshapely masses, and pyramids of unsightly rubble.[2]

It was recognised that Gloucester could suffer at the expense of Sharpness, but the general feeling was that something had to be done or trade would decline in any case. The great works at Sharpness therefore went ahead, and the new dock was opened on 25 November 1874.[3] Having bought the Worcester and Birmingham Canal in the previous year, the Canal Company now changed its name to the Sharpness New Docks and Gloucester and Birmingham Navigation Company. This was more conveniently known as the Dock Company, although the term Canal Company also continued in common usage for several years. To provide a supply of coal for bunkering steamers at Sharpness and hopefully to provide an outward cargo, the new company actively supported the construction of a railway bridge across the Severn to connect with the Forest of Dean coalfield. The bridge involved the company in considerable expenditure, but it did eventually lead to the long-awaited development of a substantial export trade in coal.

Meanwhile, two more warehouses had been built at Gloucester in anticipation of a continuing growth in the corn trade. The Alexandra Warehouse was built in 1870 for Messrs. J.E. and S.H. Fox near to the large graving dock, and the huge Llanthony Warehouse was built in 1873 for Wait James and Company just to the south of the Barge Basin. Attached to the Alexandra Warehouse was a small mill for cattle food that was operated by a twelve horse-power vertical steam engine which also worked the hoisting machinery in the warehouse. A few years after it was built, the warehouse was badly damaged by a fire which started in the roof and slowly burned its way downwards. Several manually operated fire engines attended, but their jets could not reach the upper floors. Two engines were connected to one hose which was taken up the stairs, but there were few men prepared to pump as they had not been paid enough for a previous fire, and the attempt was abandoned. One floor after another gave way, and burning corn poured out of the many windows forming great mounds around the building. After five hours, the fire had come down to the third storey where the water jets could reach, and the combined effect of eight fire engines manned by fresh volunteers gradually brought the fire under control. Many people thronged the canal banks to watch the spectacle, and in the excitement, two children fell into the canal and had to be rescued. Amazingly, much of the corn was salvaged, although most of it had been damaged by the water.[4] When the warehouse was rebuilt, the overhanging eaves where the fire had started were eliminated, and the walls were extended upwards to form parapets instead.

For a few years after the opening of the new dock at Sharpness, there was a marked increase in foreign imports reaching Gloucester. The smaller boats came up the canal as before, while cargoes from the largest ships were transferred at Sharpness to barges and lighters which

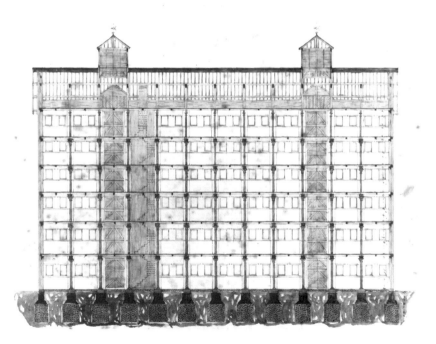

Two of the original drawings for the Llanthony Warehouse (dated 1873).

were towed up the canal by tugs. Some vessels just discharged part of their cargo at Sharpness so as to reduce their draught to suit the depth of water in the canal. The increased traffic put a great strain on the railway facilities at Gloucester and there were repeated complaints from the merchants to the two big companies serving the docks. The Dock Company had tried to improve the system by building the Victoria Bridge across the arm leading to the Victoria Dock, but they could not get the railway companies to co-operate in extending their lines. The difficulties were eased somewhat in 1876 when the Dock Company took over responsibilty for the railway lines on their land, and a further improvement was made a few years later when they built a new bridge over the lock so that they could connect the former Great Western Railway line on the west of the docks to the former Midland Railway line at the Victoria Bridge.

Facilities for the dockers and seamen were improved in 1877 by the opening of a coffee house and reading room just opposite the dock office. The project was organised by the chaplain of the Mariners Chapel, and was an attempt to provide a cheap alternative to beer. It was common for the leader of a gang of dockers to arrange with a publican to supply his men with beer in a large stone bottle, and the payment for this could become a substantial deduction from a man's wages whether he liked it or not. In due course, the coffee house was frequented by about three hundred men a day, and sandwiches and cocoa were provided as well as coffee. A rowing boat was employed to take refreshment round to the ships and to the men working in the timber yards down the canal.[5] The success of the venture led to the formation of a coffee house company that opened other branches in the town.

The period of prosperity after the opening of the new dock at Sharpness did not last long, as developments at neighbouring ports led to a spell of intense competition. The opening of new docks at Avonmouth in 1878 and at Portishead in 1880 had a particularly damaging effect, as their good railway communications enabled them to compete with Gloucester in serving the Midlands. Gloucester suffered more than Sharpness and there was even some rivalry between the two. Warehouses were built at Sharpness, and an increasing proportion of cargoes were transhipped there with some goods being transferred direct to the railways. The reduced need for handling and storage at Gloucester affected the employment of dockers and also depressed property values. Under these circumstances, the directors of the Company were particularly pleased to welcome a completely new source of traffic arising from the import of petroleum products from

Vessels crowding the quays on the approach to Llanthony Bridge in the 1880s. The schooner on the left is the Sarah Williams of Port Madoc.

America. In 1881, Messrs. Colthurst and Harding converted the former Droitwich Salt Company's warehouse at Hempsted into a petroleum store, and barrels of naphtha and benzoline were brought up from Bristol in sailing barges and trows. In the following year Francis Fox built a store just to the north of the entrance to the Victoria Dock, and he started importing petroleum products direct to Gloucester. The Dock Company were responsible for issuing licences for these stores, and they insisted on special precautions such as raised door sills, to minimise the risk of leakage. Much of the oil was sold for domestic lighting, and a shop in Southgate Street advertised 'American Standard Water White Petroleum' as a pure and safe burning oil that was imported directly into Gloucester Docks.[6] To the north of the petroleum store at Hempsted, new quays were formed to serve new timber yards, and the piling in front of the existing yards was improved to suit locomotive traffic.

The worst of the intense competition was ended by an agreement with the Bristol dock companies in 1882, and there was an increase in traffic again. Great ships bringing wheat from Australia came right up to Gloucester, and they made a marvellous sight with their masts towering above the warehouses.[7] They had to unload part of their cargo into lighters at Sharpness, and the amount discharged had to be chosen carefully to suit the depth of water in the canal without incurring unnecessary lighterage charges. However, the continuing increase in the size of merchant ships, particularly steamers, meant that access to Sharpness became a limitation. Apart from the general difficulties of navigating the river, the largest ships could only enter the dock at Sharpness on the highest spring tides a few days each month. The resulting delays and the low chance of a return cargo meant that ship owners quoted higher freight charges to Gloucester than to neighbouring ports. In response to these difficulties, there were proposals to construct a further new dock at Shepperdine (a few miles below Sharpness where there would be sufficient water to admit large steamers on every tide) and to widen and straighten the canal to increase the number of ships coming up to Gloucester. There was also a proposal to improve the Worcester and Birmingham Canal to accommodate barges carrying up to 200 tons. With the Dock Company short of money, Gloucester Corporation was urged to buy them out and to carry out the works in the interests of the city.[8] Nothing came of these ambitious schemes, however, although they continued to be talked about in one form or another for many years to come.

In the middle of all the discussions, the Dock Company's engineer, W.B. Clegram, was forced to resign in 1885 due to ill health. As clerk and then as engineer, he had given invaluable service for over fifty-five years, and in particular he had carried the main responsibility for the design and construction of the new dock at Sharpness. He was described as a careful methodical man who conveyed the impression that any statement he made was the result of calm consideration and was therefore to be relied upon. He was also a keen churchman, and had been one of the founders and chief benefactors of the Mariners Chapel.[9] To replace him, the Dock Company appointed F.A. Jones as resident engineer, and they arranged for overall supervision to be provided by G.W. Keeling who had been responsible for the Severn Railway Bridge. At the same time, they created a new post of traffic manager, and they appointed John Dixon who had been very successful in building up trade at the rival port of Avonmouth.

Dixon quickly produced a comprehensive report suggesting developments that were needed to help increase traffic. His recommendations

Looking south across the Main Basin in 1883. The brig Dr.Witte in the centre had brought 1880 quarters of wheat from Rostock in Germany.

included ways of improving the access to Sharpness, deepening the canal and docks, easing two of the worst bends, providing more storage for corn with mechanised handling and providing better towing and lighterage facilities. Although the Dock Company had little money for implementing such changes in the short term, the report provided a good basis for the future prosperity of the port, and many of the recommendations were eventually carried out at least in part. A more immediate development that Dixon was associated with was the introduction of a regular line of steamers to Antwerp and Rotterdam in 1885. This was operated by the Bristol Steam Navigation Company who had been running a similar service from Bristol for several years. To provide a place where the vessels could discharge their cargoes under cover, the Dock Company extended the roof of the iron shed beside the Victoria Dock. Gas lighting was also installed so that loading and

Steamers of the Bristol Steam Navigation Company mix with the sailing vessels beside Llanthony Quay (c. 1895).

unloading could continue at night, thus ensuring a quick turn-round. This did not prove very satisfactory, however, as there was only room to unload one boat at a time, the iron shed did not afford much storage space and the entrance to the Victoria Dock was rather narrow. The larger steamers therefore tended to use Llanthony Quay, and a new transit shed was erected there. After a slow start, traffic built up steadily, and the service was extended to include Hamburg. The main import became sugar, but the steamers also brought a wide range of other cargoes which were a valuable supplement to the traditional imports of corn and timber. The only drawback was that some of the steamers were close to the limit of what could be brought up the canal, and there were sometimes difficulties in turning them in the Main Basin if there were other vessels unloading. On one occasion, the steamship *Clio* could only be swung by opening the gates of the old graving dock.

Trading conditions remained difficult during the mid 1880s with the Gloucester corn trade being particularly affected as regular steamers brought increasing amounts of American wheat and flour to Avonmouth. Some Gloucester merchants transferred part of their business to Bristol and others went bankrupt. Even the great firm of J. and C. Sturge was forced to give up after having dominated the corn trade at Gloucester for over fifty years. Another firm to close was Reynolds and Allen, who had operated the City Flour Mills for many years. They were succeeded by Priday Metford and Company in 1886, but two years later a serious fire badly damaged their warehouse which also contained the wheat cleaning machinery and the two steam engines which powered the mill. The fire burned for two hours, the floors collapsed and a strong southerly wind almost blew the flames across Commercial Road, but the efforts of the local fire brigades managed to prevent much damage to the mill itself. At the height of the fire, the flagstaff on top of the warehouse fell sideways and plunged through the roof of the neighbouring office.[10] The fire was a serious set-back to the new firm, but the warehouse was well insured, and during the rebuilding, the opportunity was taken to incorporate silos for storing the wheat more conveniently in bulk.

The City Flour Mills (c. 1900). The original mill is on the right and the large building in the centre is the warehouse that was rebuilt after the fire in 1888.

Three pictures taken on the same day *circa* 1887. The pole and tackle in the upper picture were being used to help transfer the heavy sacks on to the cart. The horses in the lower picture were waiting to pull the railway wagon, and the crane had evidently been used for unloading stone.

Trading conditions were easier by the end of the decade, and business began to pick up again. A new processing industry was established in the docks when Fox Clinch and Company built a malt-house between the Alexandra and the Great Western Warehouses in 1888. The malt-house had four kilns with wire drying floors, and the hoisting machinery was powered by the steam engine at the back of the Alexandra Warehouse. Another new development was a tarpaulin works built by the Great Western Railway Company on their goods yard by Llanthony Quay. This building incorporated a water tower which was fed from the canal.

The recovery in trade was unfortunately disturbed by a series of labour disputes during 1889-90. Many of the labourers of Gloucester had just joined the dockers union, and they were persuaded to come out in sympathy with the men at Bristol who were trying to get a ban on the crews of foreign vessels discharging cargoes. This dispute was hardly relevant to Gloucester where most of the vessels were British, but it developed into a series of stoppages caused by union men refusing to work with non-union men. None of the strikes lasted long, but there was intermittent friction for almost a year and occasionally feelings ran very high. At least one non-union man was assaulted, bound with a rope and threatened with being thrown in the canal. The employers agreed to an increase in wages, but they insisted on their right to employ non-union labour, and more than once it required the intervention of the union's national organisers to get the men back to work.[11]

Throughout the 1880s, the Dock Company had sought ways of cutting costs and improving efficiency, and one topic which caused much heartsearching was what to do about the engineering workshops at Saul Lodge, near Frampton-on-Severn. These had originally been sited about half way along the canal to be within easy reach of all parts, but the premises were in need of modernisation and suffered from lack of any railway connection. The directors wanted to sell Saul Lodge, move the workshops to Gloucester and reduce the number of staff. The two engineers opposed this, with Keeling favouring a move to Sharpness and Jones wanting to stay at Saul Lodge. After much discussion, however, the directors confirmed their view and the move to Gloucester went ahead. A new building was therefore constructed in 1891 close to the engine house in order to accommodate a blacksmiths shop and an erecting shop, and further alterations were made over the next two years.

CHAPTER NINE

Around the Turn of the Century

By the early 1890s, there were good prospects for increasing trade, and the directors of the Dock Company were able to consider further improvements at Gloucester intended to carry them through into the next century. One such project was the development of land at Monk Meadow on the west side of the canal below Llanthony Quay. Thomas Adams and Sons of Birmingham wanted to lease seven acres for a timber yard and saw mill, and the Dock Company agreed to build a branch dock and also a quay along the canal in order to provide deep water access on two sides of the new property. To save money, consideration was given to using timber piling along the canal frontage and leaving the dock with earth sides on which wooden jetties would be constructed. Messrs. Adams were not happy with this, however, and the Dock Company eventually agreed to use concrete for all the walls. The contractor was J. Cruwys of Edgebaston, and he had one hundred men at work by the spring of 1891. He hired the Dock Company's steam crane and some iron boxes to help with the excavation, and the soil was used to build up the surrounding land which had previously been liable to flooding and was well known as a skating ground. To monitor the mixing and placing of the concrete, the Dock Company employed an inspector who had formerly been working on the Manchester Ship Canal. Completion of the work was delayed by bad weather, and in February Messrs. Adams complained that they were unable to use the saw mill that they had erected. They did not have to wait long, however, as water was let into the dock on 26 April 1892, the opening of the sluice being carried out by the chairman of the Dock Company in the presence of a score of spectators.[1] A week later, a party of shareholders was taken into the dock on a tour of inspection in the Company's steam launch *Sabrina*. All that then remained to be done was to finish clearing away the bank at the entrance and to level and ballast the wharves, and the new Monk Meadow Dock was ready for commercial use. Initially it was only served by the Great Western Railway, but several years later, the Midland Railway also obtained access by building a swing bridge across the canal.

Another improvement carried out in 1892 was the deepening of the lock between the Main Basin and the river. This was a joint exercise with the Severn Commission, and was part of a general scheme for improving the river to meet increasing competition from the railways. The original lock was in two halves with the bottom of one chamber two feet below the other, and the plan was to make one large lock that would be two feet deeper than the existing lower chamber. The Severn Commission did not want to pay for lowering the bottom of the whole lock but only a length of 150 feet, this being the length of their Diglis Lock at Worcester. Only one contractor put in a tender, and this was not thought to be satisfactory as he had not visited the site and he had not allowed enough for contingencies. As any failure of the contractor would badly hold up traffic, the Severn Commission agreed that the Dock Company should use their own men to do the work. The plan was first to close the lock completely for about two weeks so that a dam could be constructed and the flood gates at the river end could be deepened. Then the remainder of the work was intended to be carried out behind the protection of the floodgates, which would be opened for

Inside the dam at the river end of the lock (1892).

Deepening the lock in 1892.

six hours each day to allow traffic to move. This approach had to be
abandoned, however, when difficulties were experienced in making a
watertight dam. Two rows of piles were driven into the river bed with
clay in between, but too much water seeped through the earth under
the dam and a second dam had to be built closer to the lock chamber.
While this was going on, the floodgates were kept closed to allow work
on the lock itself to go ahead. Here it was found that the stone inverts
and the timber gate sills had been well made and they took longer to
remove than expected, but with gangs of men working in shifts round
the clock, the whole operation was completed in just over a month.
During this period, traffic wanting to pass to or from the river had to go
round by the Stroudwater Canal entrance at Framilode.

With imports rising again, the practice of leaving timber floating in the canal was becoming a nuisance. Large balks of timber needed to be kept in water to avoid them drying out and cracking, and so the Dock Company decided to form a special timber pond linked to the canal to the south of the new Monk Meadow Dock. An area of four acres was excavated to a depth of three feet which was needed to ensure that rafts of oak and birch could be completely submerged. The work was carried out during the first half of 1896, and the total area was divided up between several merchants who paid rents according to the space they occupied. To ensure that the new pond was properly used, the charge for floating timber in the canal was increased, and the pond was soon reported to be full. A steam crane was installed on piles near the entrance in order to assist in loading the timber into canal boats.

As well as the Dock Company's improvements during the 1890s, there were also two private developments on Bakers Quay after Price Walker and Company vacated their extensive timber yards and moved to new premises further down the canal. In 1897, Sessions and Sons built a works for the manufacture of chimney pieces from enamelled slate, a popular imitation of marble, and they used their canal-side site to import slate direct from Wales, Italy and Spain. Then two years later, G. & W.E. Downing and Company of Birmingham built a large malt-house with its upper floors projecting forward and supported on pillars in accordance with the old agreement concerning the original construction of the quay. This gave Downings direct access to the canal for the import of corn, and they also obtained permission to use water from the canal for steeping and sprinkling the grain during the malting process. A little further south, a small timber pond was constructed in 1899 on the premises of the Gloucester Railway Carriage and Wagon Works. This provided temporary storage for timber floated in from the canal, and part of the pond extended under a long gantry crane which was used to lift out the timber and carry it to where it was needed.

The railway just to the south of Bakers Quay was the scene of a tragic accident in 1898. A locomotive was pulling about thirty wagons loaded with timber towards High Orchard, and as it passed over a set of points, it lurched sideways towards the canal. The driver, the fireman and two shunters all hurriedly jumped clear, but in the confusion and darkness of the night, they did not realise that they were jumping into the path of the wagons that were still careering forward. One man was killed outright, two suffered terrible injuries and died later in hospital, and only the driver survived. At the inquest, it was suggested that the derailment might have been due to the points being misaligned because a stone had become lodged in them.[2]

Timber yards by the canal with the Monk Meadow timber pond in the distance (*c*. 1900).

Below. Construction of the Wagon Works timber pond using concrete (1899).

By the end of the century, barques of 800 tons register or more were coming up to Gloucester after unloading some of their cargo at Sharpness. Behind the barque can be seen the Great Western and Alexandra Warehouses (c. 1895).

Above right. Two large barques in the Main Basin (c. 1900). The one on the left is believed to be the *Myrtle Holm*, the largest sailing vessel to come up to Gloucester.

Right. The Dock Company's tugs *Speedwell*, *Hazel* and *Mayflower* in their usual corner beside Llanthony Bridge (c. 1895). In conjunction with *Moss Rose*, *Violet* and *Myrtle*, they provided a regular towing service on the canal and out in the estuary. The awnings were to give weather protection for the crew, but they were liable to blow away in a strong wind!

A very unusual boat arrived at the docks in August 1899. It was a thirty foot long cutter, which had been sailed single handed across the Atlantic from Gloucester Massachusetts. Howard Blackburn had spent sixty-one days at sea, mostly with light winds, and his main problem had been a badly swollen leg which had prevented him setting sail for eight days. His achievement was the more remarkable as he had previously lost all his fingers through frost-bite and his feet were also affected. Captain Blackburn was welcomed at the quayside by an enthusiastic crowd of sightseers who had to be restrained by the police, and he was then taken in an open carriage to the Guildhall for an official reception. During his stay in Gloucester, the intrepid sailor was entertained by several of the leading citizens, and he attended a performance at the Theatre Royal where he occupied a private box. Several hundred people visited the docks to see his tiny craft, and many were welcomed on board and were offered bourbon whisky and ships biscuits.[3]

Another unusual arrival was the barque *Success*, which came to Gloucester in 1906. She was fitted out like the ships that had taken convicts to Australia, and she was on a tour round the coast as a floating exhibition. Down below, there were about seventy cells like wooden cages, with wax dummies to represent the prisoners. There was a collection of prison relics on show, including leg-irons, hand-cuffs and the dreaded cat-o'-nine-tails, and there was a plaque on the floor showing where an officer was supposed to have been murdered by a convict. Along the centre-line of the ship were two ropes, one ankle-high and the other shoulder-high and visitors were told that the convicts had been tied to them as a form of punishment. In fact the *Success* had never been used for transporting convicts, but she had served as a prison ship in Australia for a number of years after her crew had deserted to try their luck in searching for gold, and there is no doubt she was a very popular exhibition.[4]

The more normal sailing vessels using the docks in the early years of this century were schooners, ketches and trows mainly trading with ports in the West Country, South Wales and Ireland. These arrived with cargoes such as stone, coal, sand and corn, and they often went out with salt. Other regular visitors were the steamers of the Bristol Steam Navigation Company which carried many loads of sugar and other goods from Hamburg and Antwerp. Smaller steamers brought linseed and cotton seed to Foster Brothers mill, and cargoes such as cement, granite sets, glass sand and naphtha came from other British and continental ports. Some steamers and large sailing vessels still brought grain and timber from the Baltic and Scandanavia, but the majority of these

The boat in which Captain
Blackburn sailed single handed
across the Atlantic, and (right)
the intrepid sailor with some
admirers (1899)

cargoes were transhipped at Sharpness and large quantities were carried up the canal in barges and lighters belonging to Mousell Chadborn and Company and G.T. Beard.[5]

The majority of the buildings around the docks were still devoted to the corn trade, including about twenty large warehouses. Most of these were run by corn merchants for their own use, but some were taken over by specialist warehousing firms who then rented space to the merchants as they required it. One large warehouse was occupied by Gopsill Brown and Sons who hired out the sacks that were needed to handle and store the corn when it arrived in bulk. Several single storey warehouses also contained corn or sacks, and some merchants had their own drying kilns. Two large warehouses were used for storing sugar, and other smaller warehouses contained petroleum products and general cargoes. Near the graving docks and along the Severn Bank, there were several workshops used by the shipwrights and smiths who serviced any ships needing repairs, and one former warehouse had become a foundry. The Severn and Canal Carrying Company had their base beside the Barge Arm or Old Arm (as the Barge Basin had become known), and a smaller firm of carriers used the adjoining premises known as St. Owens Wharf. Nearby, there were yards used by coal, stone and builders merchants, and J. Romans and Company had a large timber yard adjoining Llanthony Road. Most of the other timber yards were down the east bank of the canal, occupying a continuous frontage of three-quarters of a mile from the Wagon Works to the former salt stores near Hempsted Bridge. There were also some timber yards around Monk Meadow Dock.[6]

For some time there had been a fear that if a fire started in one of the timber yards when there was a strong south-westerly wind blowing, it could spread right through the city. This fear was reinforced in 1904 when there was a serious fire at Thomas Adam's yard at Monk Meadow, and the fire brigades were hampered by a difficulty in getting water from the mains. Fortunately, this fire was brought under control, but it was decided that a fire-float was needed to assist in any future emergency. The Chamber of Commerce raised the necessary money from the merchants with contributions from the Dock Company and the Corporation, and a vessel was built by Abdella and Mitchell at Brimscombe. The fire pumps were made by Merryweather and Sons, and they could supply up to one thousand gallons per minute through six hoses. They could also operate underwater jets to provide pro-pulsion. The new fire-float was formally inaugurated on the 12 July 1906, when it was given the name *Salamander* after the mythical creature that was supposed to live in fire. Following the naming ceremony, it

The fire-float *Salamander* demonstrating its pumping power after its official inauguration in 1906. The 'flames' on the warehouse roof have been added to the picture later!

steamed around the Main Basin and gave a demonstration of its manœuvrability and pumping power in front of a large crowd of spectators.[7]

Within a year, the *Salamander* was called to help fight a serious fire at Nick's saw mill on the east bank of the canal. The fire spread to the huge stacks of timber in the yard, and these were soon one great mass of flames which lit up the whole neighbourhood. The local fire brigades set up four hoses spraying water from the city mains, and the *Salamander* initially concentrated on preventing the blaze spreading to the neighbouring Gloucester Joinery Company. These combined efforts brought the fire under control in about three hours, but pumping continued for a further four hours until the fire was fully extinguished. It was the early hours of the morning by the time that the *Salamander* returned to its berth in the docks, but there was to be little time for the crew to rest. Later that day, they were called out again to a fire in Joseph Griggs' timber yard a little further down the canal. This fire raged for four hours and virtually destroyed a large drying shed and its valuable contents, but the jets from the *Salamander* helped to prevent the flames reaching the nearby saw mill. The *Gloucester Journal* reported that the fire-float 'shook and shuddered with her tremendous exertions, and tongues of fire shot three or four feet from her funnel'. Concern about the causes of two serious fires on consecutive days was strengthened when a third fire was discovered at Price Walker's yard on the following day. Fortunately this was extinguished before it did any serious damage, and examination of the remains confirmed that it had been started deliberately. The police made enquiries, and a young employee at the yard eventually confessed to starting all three fires. He said that he had wanted to see the fire-float at work, and he had in fact managed to get on board the *Salamander* during the second fire. He was sentenced to six months hard labour on the understanding that when he came out he would be helped to emigrate to Canada.[8]

The *Salamander* attending the fire at Grigg's timber yard in 1907.

The *Salamander* showing off its new 'monitor' when it was handed over to the Corporation in 1909.

The cargo of the trow *Spartan* being transferred to the Severn and Canal Carrying Company longboat *Swansea* (*c.* 1900). The picture shows the characteristic open hold and D-shaped stern of the trow, although the former use of square sails had been superseded by the more efficient ketch rig.

The iron-built brigantine *Sofia* moored by the West Quay (*c.* 1905). The warehouses behind were leased for a few years by the Severn Ports Warehousing Company whose main operations were at Sharpness.

A large brig by Bakers Quay with the steam packet *Wave* just setting off for Sharpness (*c.* 1905).

Steamers of the Bristol Steam Navigation Company beside Llanthony Quay (c. 1905). The regular visitors were named *Clio, Echo, Pluto, Ino, Sappho* and *Milo*, and the larger vessels were lightened at Sharpness before coming up to Gloucester.

Above right. The steam coaster *Stock Force* of Whitehaven by Biddles Warehouse (c. 1910). To help a rapid turn-round, the cargo is being unloaded into the trows *Avon* and *Wye* as well as being transferred into the warehouse. Note the use of staging to support a platform level with the warehouse floor.

Right. Longboats being manoeuvred by the use of shaft-hooks (c. 1905). The winches on the left were used in conjunction with a pole and tackle for unloading cargoes from boats that did not have their own derricks.

The cargo of the *Nelson* being transferred to the longboat *Walsall* beside Biddles and Shiptons Warehouses (*c.* 1910).

The largest carrying firm in the area was the Severn and Canal Carrying Company, which had been formed in 1873 by a merger between Danks and Sanders of Stourport and the interests of J. Fellows and Company at Worcester. The company operated some former trows which were towed by tugs to and from Avonmouth and other Bristol Channel ports, and they had a large fleet of longboats which distributed goods throughout the canal system of the Midlands. Transhipment usually took place at Gloucester, where the company occupied Biddles and Shiptons Warehouses beside the Barge Arm (or Old Arm). They also had a small steamer, the *Atalanta*, which could trade between Avonmouth and Worcester.

A group of Severn and Canal Carrying Company longboats in the Main Basin (c. 1910).

The steamer *Atalanta* (c. 1930). The uniformity of the warehouses along the West Quay shows up well in this picture.

Gloucester Docks circa 1900

Warehouses and Mills etc.
A. Great Western Warehouse
B. Fox's Malt-house
C. Alexandra Warehouse
D. St. Owens Mills
E. West Quay Warehouses
F. Lock Warehouse
G. North Warehouse
H. City Flour Mills
J. Victoria Warehouse
K. Herbert Warehouse
L. Kimberley Warehouse
M. Phillpotts Warehouse
N. Britannia Warehouse
P. Albert Flour Mills
Q. Vinings Warehouse
R. Sturges Warehouse
S. Biddles Warehouse
T. Shiptons Warehouse
U. Llanthony Warehouse
V. Pillar Warehouse
W. Downings Malt-house
X. Foster Brothers Oil and Cake Mill

Other Buildings etc.
1. Great Western Railway Transit Shed
2. Great Western Railway Tarpaulin Works
3. Dock Company Steam Towing Office
4. Severn Bank Warehouse
5. Dock Company Engineering Shops
6. Dock Company Engine House
7. Steam Packet Office
8. Lock-keepers House
9. Dock Company Office
10. Custom House
11. Corn Sheds etc.
12. Petroleum Stores (F. Fox)
13. Transit Shed
14. Albion Wharf (Sessions and Sons)
15. Mariners Chapel
16. St. Owens Wharf (Jacob Rice and Son)
17. Victoria Wharf (John Knight)
18. Romans Timber Yard
19. Bridge-keepers House
20. Slate Works (Sessions and Sons)
21. Former Tar Shed (Bristol Steam Navigation Co.)
22. Midland Railway Transit Shed

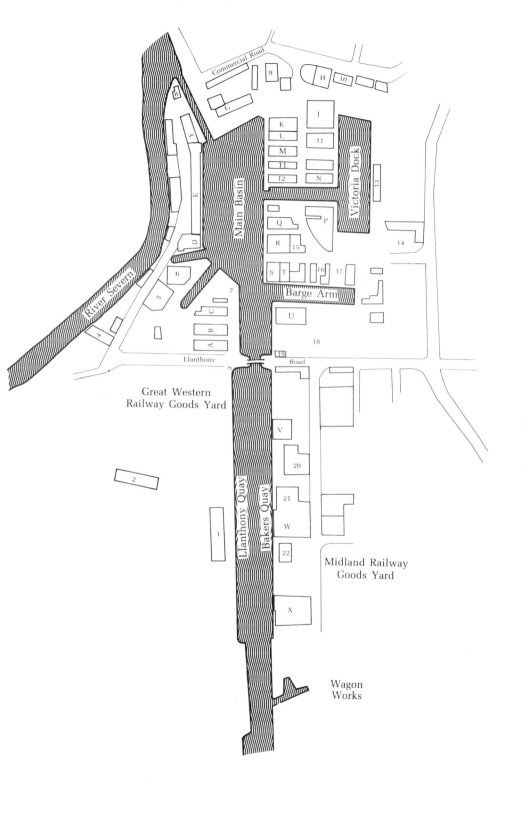

Commercial Road

9

H 10

G

K
L J
M 11
11.
12 N

Main Basin

Victoria Dock

Q P
R 15

S T 16 17

Barge Arm

U

14

18

19 Road

River Severn

D

6

5

7

C
A B

Llanthony

Great Western
Railway Goods Yard

2

Llanthony Quay

Bakers Quay

V

20

21

W

22

Midland Railway
Goods Yard

1

X

Wagon
Works

CHAPTER TEN

Cargo Handling

In the early years of this century, cargo handling was still mainly done by manual labour with a few simple mechanical aids. As payment was usually by piece-work, highly efficient techniques were developed involving considerable teamwork. Grain usually arrived in bulk and had to be sacked-up for the transfer to longboats or to a warehouse. Each sack was filled by a man using a vertical-sided metal bucket known as a bushel. (Originally the grain was measured by volume and a sack contained four bushels). The bushel was pushed into the grain with the knees, flipped upright, topped up with two scoops of the hands and then emptied into the sack. When the sack was full, a self-tightening loop on the end of a hoisting rope was thrown around its neck. This had to be done smartly as any delay could result in the man getting his hand caught in the loop and being lifted up with the sack. For unloading sailing ships, the hoisting rope was slung from a derrick rigged from one of the masts. Each sack was first hoisted up on the deck where the contents were adjusted to the correct weight, and then it was lowered over the side either into a waiting longboat or on to the quay. Sacks unloaded on to the quay were lifted into a warehouse using a rope from a projecting gable and a winch worked by two men in the warehouse loft. The rope was sometimes arranged so that as one end was being used to lift a sack, the other end was descending ready for the next sack. To unload lighters and barges which did not have a mast on which to mount a derrick, the warehouse hoist was used to give a direct lift in spite of the awkward angle due to the warehouse being set back from the quay. In this case, planks were leant against the warehouse wall to break the swing of the sack as it was lifted clear of the boat. In spite of this precaution, occasional accidents occurred, and it was usual to lay a canvas sheet on the quayside to catch the corn from any bag that split.

Once in the warehouse, the sacks were stacked up two or three tiers high. The lower sacks could be wheeled into position using trucks, but the upper ones had to be carried by the dockers on their shoulders even though each sack weighed over two hundredweight (100 kg.). It was hard work moving the heavy sacks about all day, but the men became highly proficient in handling them and could usually put them down in

Corn porters with their sack trucks (c. 1900).

just the right place to ensure the stability of the stack. Alternatively, the grain was stored in bulk and the sacks were returned to the boat to be used again. The sacks were provided by specialist firms who hired them out by the day and ensured that they were cleaned and mended if necessary ready for re-use. When the corn was taken out of store, all the operations were repeated in reverse order. Breaking down a high pile of sacks was usually done with care to avoid any risk of collapse, but occasional accidents did happen and at least one labourer was killed when some falling sacks knocked him down and crushed him. When a warehouse was being emptied, a rat hunt was usually organised, and men were employed to kill the rats by kicking them or hitting them with large pieces of wood. Good ratting dogs were also welcome, and great rivalry developed as to whose dog would claim the most victims.

113

A gang of corn porters with their tools (c. 1900). Some of the men are sitting on the bushels used for filling the sacks, and the man in the middle has the scoop for topping up the sacks to the correct weight. The dogs were for catching the rats that inevitably accompanied corn shipments.

Above right. The cargo of the barque *Gers* being unloaded with the help of a steam winch (c. 1900). The winch is mounted on an old barge which is held away from the barque by balks of timber so that a longboat can come in between to pick up some of the cargo.

Right. The winch in the loft of the Britannia Warehouse (1982). Two men could stand on the platform to turn the handle. The brake lever was operated by a rope hanging outside the building and controlled by a 'hatchman' on the floor into which goods were being loaded.

Unloading a timber lighter at Price Walkers yard (c. 1930).

For unloading timber, lines of planks extending as much as a hundred yards from the boat were set up on trestles for the dockers to run along carrying the wood in a highly organised operation. Men working in the boat 'boxed up' similar sized pieces to make a 'handful' weighing up to one hundredweight. This was 'thrown up' by two men on to the shoulder of a carrier who required considerable experience to judge the balance point correctly or otherwise he could drop the load and so waste time. The carriers wore special leather pads to protect their shoulders, and the load was steadied with one hand on top to hold the front end tipped down slightly. In this way they managed to balance and to turn great lengths of wood with apparent ease, and their boast was 'what two men can lift, I can carry'. They ran along the planks to where their particular size of wood was being stacked, 'flopped the handful' down on top of the pile and ran back to the boat for more. As the piles of timber grew, the access planks were raised high above the ground, and the carriers needed great skill to compensate for the natural movement of the narrow swaying planks as they hurried over them. The construction of each pile was closely super-

116

Note the planks set up on trestles for the carriers to run along.

vised to ensure that it remained stable to its full height and that there were adequate air passages to promote drying. This was carried out by a man at one end of the pile who adjusted the position of each piece of wood delivered by the carriers. As the pile grew, he left occasional 'horns' of wood protruding so that he could lay a plank across two horns to stand on. Speed was important to avoid delaying the ship that had brought the timber, and the men were paid a piece-work rate for the amount that they moved. A gang comprised about twenty men who usually worked hard and with great efficiency. They stopped occasionally for refreshment which often included a drink of beer. During the importing season, they could earn very good money, but a proportion was deducted by the gang leader to pay for the beer. When the wood was required for use, it all had to be handled again. Individual pieces were pushed over the end of the stack so that they stood upright supported by the horns protruding from the stack. Then the pieces were carried to railway wagons ready for dispatch. Speed was not so important for this operation, and it was done by a gang of five or six men paid by the day.

117

Deals being unloaded from a steamer at Romans yard (c. 1925). Bundles of wood are being lifted ashore by the ship's derricks, and the men are carrying pieces to the appropriate stack in the yard. Other men are supervising the construction of the stacks, while the carriers return along the planks in the middle of the picture to collect more.

Above right. Deals being unloaded from a lighter in the Barge Arm (c. 1935). The carrier is steadying his load before starting his run along the planks, and another man is preparing to box up a load for the next carrier.

Right. A gang of timber yard men with their leather shoulder pads (c. 1910). Note the large bottles for their communal supply of 'refreshment'.

A vessel calling to pick up salt usually moored in the Victoria Dock, and a longboat that had loaded at Stoke Prior moored alongside. The transfer was made by shovelling the salt into a large basket which was then lifted by a derrick and emptied into the hold of the visiting vessel. Some salt came by rail, and then it was shovelled out of the wagon on to a large chute running down into the hold of the vessel. Salt was not normally stored at Gloucester, but small amounts left over were occasionally kept in the transit shed to the east of the dock ready to be put on board the next boat to call. Coal was brought by long boats to the Barge Arm for local distribution, and again baskets were used for unloading at least the smaller pieces. The larger lumps were put in a wooden cradle which had handles at each end so that it could be carried by two men.

For heavy loads, there were some old hand-operated cranes of various sizes, but these were not used much as they were rather slow. They had been largely superseded by a number of mobile steam cranes that could be trundled round the dock railway system to where they were needed. They were particularly used on Llanthony Quay for unloading the regular steamers belonging to the Bristol Steam Navigation Company, and many different commodities were quickly transferred to waiting railway wagons or to temporary storage in the transit shed. Large quantities of sugar were handled in this way, and many wagonloads of sacks and crates were taken round to the nearby Great

Coal being unloaded at the Barge Arm (c. 1905).

120

Longboats loaded with salt in the Victoria Dock which was also known as the Salt Basin (c. 1895).

A hand-operated crane beside the Barge Arm being used to load newsprint in 1926. The posts of this crane and of the one behind still survive.

The Main Basin before the First World War.

Western and Alexandra Warehouses to await distribution. The steam cranes were occasionally used at the timber pond to raise large balks from the water, and they were also used for other heavy lifts such as cement and roadstone etc. In 1909, a new boiler was being transferred from a railway wagon to a barge on its way to Butlers tar works at Sandhurst, and spectators were horrified to see the crane and its load tip over into the water, tearing up part of the railway line as it went. The cabin of the crane was smashed, but the driver just managed to extricate himself in time and he swam to safety.[1] The load should have been well within the capability of the crane, but it was noticed that the boiler was only just floating, and subsequent investigation showed that a lot of water had been left inside by the manufacturer so that the boiler weighed twice its proper weight. In Price Walker's timber yard down the canal, there was a steam-powered gantry crane which was used for handling heavy logs. Water for the steam engine on the traveller was pumped up to a holding tank by a wind-powered pump, the tower of which formed a prominent landmark.

Horses were used on the dock railways to distribute incoming wagons, marshall outgoing wagons and make short transfers between quay and warehouse etc. These movements could be quite time-consuming as many of the lines along the quays were only linked to the main system by small turn-tables that could only take one wagon at a time. The horses knew exactly where to stop to position a wagon correctly on a turntable, which was then rotated by pushing on an extending arm. Sometimes a horse was hitched up to the arm to get the turntable moving, but care was required not to exert too much force or the wagon could swing round out of control. The horses also knew that it was worth stopping outside the black shed beside the Great Western Warehouse, as this was where the lump sugar was stored and some-body would usually give them a lump or two. Locomotives were used for transfers between the different parts of the docks, and as some of the bends were rather sharp, the Midland Railway had a special type of tank locomotive with a very short wheelbase.

The *Valdivia* brought 5,000 quarters of wheat from Australia in 1913. She was probably the last barque to come to Gloucester.

In 1913, the Severn and Canal Carrying Company brought into service the motor barges *Osric* and *Serlo*, which could carry their own cargoes and also tow dumb-barges and longboats. In the picture above, the *Osric* is setting out for Sharpness towing the *Togo*, while a number of lighters and longboats are waiting for a tow from the tug *Moss Rose*.

Above Right. The *Osric* and *Togo* approaching the Wagon Works and Monk Meadow.

Right. The chaplain of the Mariners Chapel preparing for an open-air service (c. 1915). The small hand-cart was used to move the organ which can be seen on board the *Osric*.

CHAPTER ELEVEN

A Changing World

The outbreak of the First World War had a disastrous effect on the trade of the port. Half of the corn imports were lost due to the closing of the Black Sea to trade, and both the timber imports from the Baltic and the steamer trade with the Continent ceased altogether. Total imports into Gloucester and Sharpness fell from 679,000 tons in 1913 to 208,000 tons in 1918, and exports fell from 268,000 tons to 112,000 tons over the same period. The merchants struggled on as well as they could, but the pre-war pattern of life had gone forever. Towards the end of the war, a serious fire destroyed the warehouse built on pillars on the West Quay. It was thought that the blaze was started by a fault in the wiring for an electric-powered hoist that had just been installed. The fire spread rapidly upwards, and soon the whole building resembled a raging furnace. The flames started to spread to the roof of the neighbouring warehouse by the lock, but the fire-float *Salamander* was brought into action and helped to save the main structure of this building.[1]

In the final year of the war, a yard for building reinforced concrete barges was established beside the canal just to the south of Hempsted graving dock. The barges were ordered by the Admiralty as part of a national programme to save steel. They were built by a firm formed by Messrs. Hobrough and Company and Messrs. W.T. Nicholls, and Mr. W. Leah acted as constructional engineer. Each barge was 180 feet long and 30 feet wide and was designed to carry 1000 tons of cargo. They were built in pairs on the same slipway and launched broadside into the canal. The first one named *Creterock* was launched on 23 November 1918. As the barge entered the water, there were cheers from a large crowd of spectators, but then the cheers changed to laughter. The sudden entry of the huge barge into the canal created a tremendous wave which washed right across the towing path on the opposite side, and a number of people standing there were soaked.[2] After the launch, the *Creterock* was towed up to the Main Basin for fitting out. Five other barges were subsequently launched from the yard, and these were named *Creteroad*, *Creteridge*, *Creteriver*, *Creteravine* and *Creterampart*.

As part of an effort to encourage a recovery in trade after the war, the

The 1000 ton capacity concrete barge *Creteridge* launched in 1919.

Dock Company built a transit shed on the south side of Monk Meadow Dock in 1921. The shed was a former hangar obtained from Rendcomb Aerodrome, and it provided a large storage area on one level which was suitable for vehicle access. The continental trade started again, but grain and timber imports were still depressed, and traffic remained well below pre-war levels throughout the 1920s. One significant development during this period, however, was the start of a new phase in the import of petroleum products to provide for the growing number of road vehicles with internal combustion engines. Storage depots were established at Monk Meadow Dock by firms such as Anglo American, British Petroleum, National Benzole, Red Line, Russian Oil Products and Shell Mex, and coastal tankers brought in the petrol for distribution to the local area. There were also some developments in cargo handling, as suction wheat intake plants were installed at the Albert Mills and the City Flour Mills.

The suction wheat intake and conveyor for the City Mills (1925). Before this was installed, all the wheat had to be sacked-up to be taken the few yards to the mill by horse and cart.

The steam coastal tankers *Ben Read* and *Ben Robinson* in Monk Meadow Dock (c. 1925). The tall funnels and caps were to prevent sparks.

The tanker barge *Shell Mex 7* being towed through the Main Basin (c. 1940).

The locally owned schooner *Excelsior* in the Main Basin (c 1925). There was still a fair number of schooners and ketches trading during the twenties and thirties in spite of growing competition from motor vessels.

Above right. The well-known schooner *Dispatch* waiting to take on a load of salt in the Victoria Dock (c. 1930).

Right. A large queue has built up waiting to collect the cargo being unloaded by the steam crane beside the Victoria Dock (c. 1925).

In 1926, the dockers took part in the General Strike and effectively kept the docks closed for two weeks. The strike committee did arrange a system of permits to 'facilitate' the movement of foodstuffs, but they would only deal with union members, and non-union lorry drivers who called to collect sugar found their way barred by the pickets. There was further trouble when it became known that the tug *Victor* with a volunteer crew was going to take two barges loaded with corn up to Tewkesbury. A crowd of about three hundred strikers assembled by the lock, and they tried to prevent the check ropes being secured as the boats entered the river. There was a strong force of police on duty, however, and the boats got through safely, although the men on board had to suffer abuse and stone throwing as the crowd pursued them along the river bank. On the following day, the tug *Speedwell* was taking an empty barge for loading at Sharpness when it was held up by three to four hundred strikers at Hempsted Bridge. The bridge was in two halves, and although the police managed to swing open one side, the strikers kept the other side closed. The police then closed their side with a view to crossing the bridge, but the dockers managed to frustrate the plan by opening their side. After both sides of the bridge had been swung back and forth again to no effect, the stalemate was relieved when a union man on board the tug agreed to leave, and the tug was allowed to proceed on its way to Sharpness. The high feelings generated during this period were not easily pacified, and even after the national strike had collapsed, the men stayed out until a local settlement was arranged by Ernest Bevin, the union national secretary.[3]

The growing use of road vehicles had a serious effect on the steam packets that had operated between Gloucester and Sharpness for almost eighty years. The original boats had been replaced, but the names *Wave* and *Lapwing* had passed to their successors, as had their role of serving the villages along the canal. The packets stopped at the various bridges to pick up people coming to shop in Gloucester or children on their way to school, and on the return journies, parcels were often taken down the canal to be left at the bridges for collection. The boats also took parties to the Plantation pleasure gardens at Sharpness, or perhaps just to the junction with the Stroudwater Canal from where there was a short walk to the river at Framilode. They had first and second class cabins, but many passengers preferred to remain on deck to enjoy the view. During the summer, refreshments were provided on the afternoon trips, using hot water from the boiler to make the tea. With the advent of country buses, however, passengers began to forsake the *Wave* and *Lapwing* for the faster road service, and the two boats were withdrawn in the early 1930s.

The steam packet *Wave* setting off for Sharpness and passing through Frampton Bridge (early 1900s).

The tunnel tug Worcester which normally towed longboats through the tunnels on the Worcester and Birmingham Canal (c. 1930).

A motor longboat in front of Llanthony Warehouse (c. 1935).

134

Price Walker's new gantry crane in 1932. The original gantry was converted into the timber shed on the left of the picture.

Speedwell in the small graving dock (1936).

Wheat was brought to the Gloucester mills in wooden barges that had once been sailing vessels (c. 1935).

During the 1930s, traffic on the canal returned to pre-war levels, although most of the movements were barges and lighters bringing cargoes that had been transhipped at Sharpness or Avonmouth. Enormous quantities of timber were unloaded at the yards along the canal, and much was taken on up the River Severn for use in the Midlands. Wheat continued to come to the City Mills and the Albert Mills, barley came to Downings malt-house, and oil seeds were brought to Foster Brothers mill. The word soon spread if a load of peanuts or sweet locust beans arrived, and local children were quick to help themselves to a sample if they had half a chance. Competition from the railways, however, had reduced the amount of grain going further inland by boat, and less than half of the warehouses still retained their traditional role. Several of the warehouses were left vacant, others were let to firms who made little or no use of the docks themselves, while some were taken over by builders merchants who were also increasing their use of other dock premises.

Timber lighters waiting to be unloaded at Price Walkers yard (c. 1930).

Sacks of cotton seed from Bombay being unloaded at Foster Brothers mill (1937).

Copper ingots being transferred from the *Severn Trader* into a longboat (1938).

The Severn and Canal Carrying Company introduced a number of motorised barges that could operate down to Avonmouth, Bridgwater and Swansea and could also pass through the locks up to Worcester and Stourport. Cargoes for Birmingham were still usually transhipped at Gloucester. The company's longboat fleet was smaller than before the war, but was particularly used to carry cocoa beans, sugar and chocolate crumb to Cadbury's factory at Bournville. The chocolate crumb was very popular with local children and occasionally one of the bags would be 'accidentally' damaged and some of the contents 'lost in transit'.

Another view of the *Severn Trader* being unloaded.

Transhipping in the Barge Arm (*c.* 1935). The man controlling the derrick can be seen in the doorway wearing a safety harness.

139

Bags of cement being transferred into the transit shed beside the Victoria Dock (1930s).

In addition to the barge traffic, steamers brought cargoes such as cement from London, cotton seed from Hull, granite chippings from North Wales and sugar from London and Cantley (Norfolk). The new traffic in petroleum products developed rapidly, and as well as the coastal tankers coming to Monk Meadow, John Harker and other companies operated a number of tanker barges that picked up supplies at Avonmouth and delivered to depots at Worcester and Stourport. Some small sailing vessels still came to the Victoria Dock to collect loads of salt (by this time usually brought by rail) to take across to Ireland, but this was much less common than in fomer years due to competition from Liverpool.

Granite chippings being unloaded from the steamship *Kyle Firth* in 1936.

The steamship *Gorsefield* brought 356 tons of apples from Honfleur in 1938.

The Irish schooner *Brooklands* was one of the last to load salt in the Victoria Dock (*c.* 1940). She became the last trading schooner without an auxilliary engine.

A pair of submarines on a courtesy visit in 1937.

The L.M.S. tank locomotive No.1537 crossing Llanthony Bridge into G.W.R. territory (1936). Note the additional coal on top of the boiler.

During the Second World War, the west coast ports had to deal with additional ships diverted from London, and the canal and docks played a vital role in handling essential cargoes for the Midlands, particularly petroleum, metals and foodstuffs transhipped at Avonmouth. The barges usually returned empty, although there were outward cargoes of jerry cans, electrical equipment and service stores at the time of the Normandy landings. The barge crews worked all hours of the day and night to keep supplies moving, and received extra rations as they were classed as merchant seamen. Some fishing trawlers were brought to the docks to be converted into minesweepers, and several wooden harbour launches were built for the navy by a gang of carpenters with little previous experience of boat building. Some of the warehouses were brought back into use for storing corn as part of a strategic reserve, and holes with shutters were made in the floors so that corn that had been blown up to the top floor by a fan could be run down to the required level. In 1943, a grain silo was built beside the canal near Monk Meadow Dock, and this was used mainly for drying and storing the large amounts of local corn grown during the war. In the same year, the coaster *Empire Reaper* started bringing regular shipments of coal to the newly opened Castle Meads power station on the west bank of the River Severn. The coaster had to pass through the lock to reach the power station jetty, and the bed of the river was blasted and dredged to provide sufficient depth of water. There was not enough space for such a large vessel to turn round in the river, and so when the coal was unloaded, the coaster had to be manœuvred backwards into the lock with the help of wire ropes. The *Empire Reaper* brought almost four hundred tons of coal from Barry every week or two for about two years.

In January 1945, a serious fire virtually destroyed the Great Western Warehouse near Llanthony Bridge. The building was being used for the preparation of oatmeal for breakfast cereals, and about forty people were working there when a small dust explosion occurred at the base of a hopper where the oats were being ground. The fire spread quickly throughout the building and soon huge flames could be seen leaping upwards to a height of 150 feet.[4] The fire-float *Salamander* was brought across from its berth to fight the fire, but initially it was not tied up properly, and when the water jet was turned on, the whole boat swung round and the jet sprayed over spectators who had gathered on Llanthony Bridge. The fire was subsiding after an hour, but the eventual damage to the building was so bad that the upper part had to be demolished and the ground floor was converted into a single storey warehouse.

The first visit of the *Empire Reaper* in 1943.

The motor vessel Severnside discharging wheat for the Albert Mills (1953).

After the war, it was clear that the Dock Company did not have the resources to provide the investment needed for further development, and in 1948 it was absorbed by the newly formed Docks and Inland Waterways Executive which later evolved into the British Waterways Board. The new management set about encouraging sea-going ships to come up to Gloucester again. They replaced the old double swing bridges on the canal with single leaf bridges, and they took over the former Great Western Railway Company's Llanthony Quay (which was also known as the Western Wall). Here they built single storey warehouses at either end of the old transit shed to be suitable for lorries and fork-lift trucks. These moves successfully stimulated direct imports of fertilizer, foodstuffs, timber etc. and exports of home grown grain and boxed car parts for assembly in Ireland.

The motor vessel *Reginald Kearon* entering the Victoria Dock to load boxed car parts (1961).

Sacks of chocolate crumb from Cadbury's factory at Waterford being transferred to railway wagons (1954).

Wheat for the Albert Mills being discharged from the former sailing vessel *Hannah* (c. 1950). The earlier suction discharge plant had been replaced by Redlers elevators.

Preparing to launch a pontoon in 1956. It was unofficially given the name H.M.S. *At Last* because it had taken a long time to build.

The motor vessel *Ahoy* discharging bulk potash in the Victoria Dock (1964).

Petroleum traffic continued to expand, and John Harker and Company had over twenty tanker barges operating on the Severn, most of them having names ending in 'dale'. The largest ones could carry 400 tons of petroleum from Avonmouth or Llandarcy up to Worcester. Shell Mex-BP also had some barges serving Gloucester and Worcester, and Regent had several smaller boats operating up to Stourport. These movements reached a peak around 1960 when tanker barges could be seen queuing in the Main Basin, waiting their turn to pass through the lock on their way up the river. The traffic rapidly declined, however, following the construction of pipelines during the 1960s. Deliveries to Monk Meadow Dock also came to an end following the establishment of a new depot further down the canal at Quedgeley.

The 375 ton capacity tanker barge Westerndale H entering the lock (c. 1955).

The large graving dock was much used for maintenance work on the tanker barges (1960).

Arkendale H in 1960 shortly before she was involved in the tragic accident which demolished two sections of the Severn Railway Bridge.

151

A rare coaster discharging wheat from the Albert Mills – the *Sophia Weston* in 1974.

The copper being loaded on to the motor vessel *Biak* is the last known outward cargo loaded from a barge (1962).

The Dutch motor vessel *Twin* brought 690 tons of basic slag fertilizer from Dunkirk in 1960.

During the nineteen-sixties, coasters continued to use the Main Basin and Llanthony Quay, and a new deep water quay was constructed beside the canal to the south of Monk Meadow Dock in 1965. However, barge traffic declined in the face of competition from road transport and the introduction of modern handling methods. Increasingly the timber arriving at Sharpness came in bundles that could readily be carried on articulated lorries, and other cargoes too were transferred direct to lorries and so bypassed Gloucester. Even the wheat shipments to the City Flour Mills came to an end as supplies were obtained direct from Liverpool via the motorway. The Albert Mills continued to receive wheat by boat, but this traffic also ceased when the business was closed down in 1977.

CHAPTER TWELVE

The Docks Today

In spite of the decline in traffic, the old prophecy that the docks would become puddles and the warehouses fall to rubble has not come to pass. At the time of writing in 1983, most of the features described in the earlier chapters can still be seen and the area is full of interest.

Coasters continue to use the quays south of Llanthony Bridge, and they come into the Main Basin in order to turn. Also, a few motorised barges still pass through the Main Basin carrying wheat from Avonmouth to Healings Mill at Tewkesbury. As the other commercial traffic has died away, it has been partly replaced by an increase in pleasure craft, and the docks have become a popular place for moorings. In 1980, a very special boat rally was held to mark the 400th. anniversary of the charter giving Gloucester the formal status of a port. The basins were crowded with boats and there were many land-based attractions, and over 15,000 spectators visited the docks during the two days of the rally. To mark the occasion, the Duke of Gloucester unveiled a commemorative plaque on the wall of the British Waterways Board offices. The two graving docks and the adjoining engineering shop are still used for the repair of small sea-going and inland waterways craft. One boat overhauled in 1982 was the old sailing ketch *Irene*, which is now just used for cruising but had formerly been a frequent visitor to Gloucester coming to pick up cargoes of salt. Water is still pumped into the Main Basin from the River Severn, although the old steam engines have long since been replaced by electric motors and the chimney of the engine house is now only a stump. As well as providing the needs of the canal, the water is used to supply Bristol via a treatment plant at Purton. The water from the Severn has a high silt content, and this tends to settle out in the Main Basin so that a dredger must be used regularly to maintain the proper depth of water. The silt is transferred to lighters which are towed down the canal for the silt to be pumped back into the river again south of Purton. A modern dredger is now used for this duty, but the old steam-powered dredger may still be seen moored in the docks. The steam tug *Mayflower* was also moored in the docks for many years after its working life was over, but it has now been taken to Bristol Docks to become an exhibit in the museum there.

Most of the main warehouses have survived, although few are used

The Main Basin in 1983.

much because the modern preference is for single storey buildings with access for fork-lift trucks etc. Only the continued operation of the City Flour Mills is a reminder of the once great corn trade. The warehouses should continue to survive as they are all now listed by the Department of the Environment as being of special architectual and historic interest. They form a unique monument to that great Victorian era of Free Trade, when Britain imported huge quantities of corn and raw materials and exported manufactured goods all over the world. Although minor alterations and additions have been made over the years, most of the warehouses are still much the same as they were when built. The main casualty has been the row of early warehouses along the West Quay that was finally demolished in 1966. The North Warehouse is in poor condition as two of the roof tie-beams have rotted through. The associated rafters therefore give an outward thrust on the front wall which has had to be supported by scaffolding. A Public

The ketch *Irene* in the large graving dock (1982). She was built in Bridgwater in 1907 and is the last surviving West Country coasting ketch.

Pleasure craft gathering for a boat rally in 1982.

Enquiry was held in 1981 to consider a proposal to demolish the building, but permission was refused. There is an urgent need to find new uses for these old buildings, and the conversion of the Lock Warehouse into an antiques centre is an excellent example of what can be done. Additional information on the design and construction of the warehouses and on the people associated with them is given in Appendix 1.

Many of the lesser buildings have also survived. The former office of the Canal Company is now the regional office of the British Waterways Board, and the Board also occupy some former merchants offices in Commercial Road. The Custom House is now the museum of the Gloucestershire Regiment. The lock-keeper's house is still in use, as is the bridge-keeper's house by Llanthony Bridge and two other cottages by the Southgate Street entrance to the docks. The former weigh-bridge house just outside this entrance appears to be a copy of the old bridge-keeper's houses along the canal. The Mariners Chapel is still in regular use, although few mariners are now to be seen in the congregation. There are also some large single storey buildings that used to be known as corn sheds, some smaller storage buildings, and several offices built for the merchants and their clerks that are mainly clustered around the Commercial Road entrance.

The Victoria Dock in 1981 with the Britannia Warehouse to the right of the Albert Warehouse.

Greater changes have taken place to the south of Llanthony Bridge, but there is still much of interest to be seen. Several of the former timber yards have now become the sites for modern factories, but some of the old yards are still operating and new ones have been formed. Although most of the wood now arrives by road, J. Romans and Company have started using small coasters to import cargoes direct from the Baltic again. The railway network has been reduced drastically, and the only remains of the former Midland Railway goods yard at High Orchard is the iron transit shed by the canal. Most of the sidings have also been removed from the Great Western Railway yard by Llanthony Quay, but a rail connection still remains to serve the quay and a cement depot in the yard, and it also extends to the grain silo further down the canal. On Bakers Quay, the Pillar Warehouse is being converted into a restaurant and wine bar, the former slate works is used for making plastic mouldings and Downings malt-house and Foster Brothers mill are both occupied by West Midlands Farmers. There are still oil storage facilities around Monk Meadow Dock, although they are no longer supplied by boat. The transit shed to the south of the dock has been modernised, but its original wooden structure can still be seen inside. The timber pond to the south is now filled in, and only the entrance from the canal is recognisable. On the other side of the canal

Monk Meadow Dock in 1973 showing some of the oil installations. Also in the picture is the grain silo, and across the canal is the former Wagon Works.

has survived a small section of the gantry crane which replaced the earlier one on the same site, but the timber yard it once served is now occupied by a factory. Further down near Hempsted Bridge, one of the old salt warehouses is used for storing timber and the other for making corrugated cases. Below the bridge, the small graving dock is still used occasionally for boat repairs and maintenance, and nearby the careful observer will find the remains of the ramps that were used for launching the concrete barges at the end of the First World War. Finally, some traces can be seen of the timber pond at the Two-Mile Bend that was used by Morelands for storing the logs that they made into match splints.

Although large changes have taken place to the south of Llanthony Bridge, the appearance of the main docks area has altered so little over the past hundred years that it makes an ideal location for filming historical drama. Many scenes for the popular television series 'The Onedin Line' have been recorded there, particularly in front of Biddles Warehouse. In 1982, the square rigged ships *Soren Larsen* and *Marques* visited the docks to take part in programmes about the life of the explorer Shackleton and about an early German railway engineer. With actors in period costumes and a variety of horses, wagons and suitable props, the film crews managed to re-create briefly the bustle of activity that must once have been common. The docks have also become a tourist attraction, with organised parties being shown round by guides from the Gloucester Civic Trust and with boat trips down the canal and up the river on the *Gloucester Packet*. Development proposals are likely to bring some changes in the future but hopefully these will take full account of the unique architectural and historic setting, and will in turn benefit from it.

The *Marques* taking part in filming for a German television programme in 1982. The single storey 'building' in front of the warehouse was specially constructed for the occasion.

Filming a scene for 'The Onedin Line' in 1978.

The Warehouses

The main warehouses at the docks are all remarkably similar, considering that they were built over a period of almost fifty years. They all have brick walls, slate roofs and timber floors supported by cast iron columns, and almost all are set back from the quayside. They have many small windows, usually with stone lintels and cills. They have loading doors on each floor, and a manually operated winch was originally installed above each set of doors. Access to each floor is via a wooden spiral staircase formed round a newel post. This uniformity of style was partly due to the controlling influence of the Canal Company and partly because almost all the warehouses were built for the same purpose (i.e. the storage of corn). The Canal Company exercised their influence by imposing conditions in the leases of the plots of land on which warehouses were to be built, and in the early days at least, they had a deliberate policy of ensuring a reasonable uniformity of structure and appearance. The buildings were set back from the water to allow general access to the quays which remained the property of the Canal Company. For storing corn, the buildings needed to accommodate high floor loadings and to have good ventilation. To carry the high loads, the floors rest on massive wooden beams, typically thirteen inches square spanning the whole width of the building. These are supported on brick piers at each end and on intermediate hollow cast-iron columns. To avoid the beams being crushed locally by loads from the columns above, solid cast-iron pins pass through holes in the beams and transmit the vertical loads directly to the columns below. Ventilation was provided by window openings which originally had wooden shutters on the inside rather than glass. At least on the ground floor, the openings were fitted with vertical iron bars in order to improve security and to comply with the Customs requirements for bonded stores. The interior walls were usually coated with whitewash to minimise the need for artificial illumination. The Canal Company were very concerned about the fire risk, and they prohibited the use of naked lights. They also discouraged the use of open fires, but some warehouses were built with a fireplace in one corner of the ground floor to heat an office. These offices had slightly larger windows than the rest of the building.

Although the warehouses are so similar, some evolution in detailed

The interior of the Llanthony Warehouse.

design can be observed. Particularly noticeable is the trend towards larger size, the change from facing the dock to being end-on to the dock and the change from brick vaulted basements to foundations in the form of inverted brick arches. The early warehouses (mostly now demolished) were around sixty feet long parallel to the dock and thirty feet wide with three or four stories. The lowest floor was a few feet above the ground at a level suitable for loading to and from horse-drawn carts, and a brick vaulted basement below had its own entrance and so could be sub-let as a separate unit. Vertical iron bars were fitted to the window openings of all floors (or almost all). The two halves of the North Warehouse are the only surviving examples of this early style, although Biddles Warehouse and the Lock Warehouse show some of the features. During the 1830s, a number of experiments were tried and a new general style evolved. The later warehouses are around one hundred feet long and forty feet wide with six stories and a loft, and with the gable end facing the dock to make better use of the available frontage. The lowest floor is at or just below the ground level, and the foundations often include inverted brick arches to spread the load transmitted down through the walls and the cast-iron columns. Iron bars are usually only fitted to the window openings on the ground floor and occasionally the first floor. The storage capacity is about 10,000 quarters of wheat weighing about 2,250 tons. Most of the surviving warehouses are of this general style, although there are some variations and three are effectively double units side-by-side.

Most of the early warehouses were erected for individual merchants who wanted to use the finished buildings for their own trade. Later, it became common for financiers to arrange the construction and then sub-let to a merchant. The lease for the land was usually for a period of sixty three years, after which the building became the property of the Canal Company. Several local architects provided designs, and many of the original drawings are preserved in the Gloucestershire Record Office. Best known of the architects was S.W. Daukes, who designed Sturges Warehouse and probably the Pillar Warehouse. John Jacques designed Phillpotts Warehouse and probably several of the others that are virtually identical. Capel N. Tripp designed the Llanthony Warehouse. Construction work was carried out by local builders and seems to have usually taken six months to a year. By the end of the nineteenth century, it had become common for occupiers to paint their names in large letters on the outside of the warehouses. Most of the early signs have now been painted over by later occupiers, but some old and faded lettering can still be deciphered. Several of the warehouses are still known by the name of the original owner, but where the owner had several warehouses, other distinguishing names have come into common usage.

More information on the surviving buildings and on the people associated with them is given in the following schedule.

NORTH WAREHOUSE

Alternative name	Canal Company's Warehouses
Year built	1826-7
Original owners	The Canal Company
Architect	Bartin Haigh of Liverpool
Builder	William Rees and Son
Users	Many different corn merchants.

A stone tablet just below the cornice has the inscription 'The Glocester and Berkeley Canal Company's Warehouses Erected by W. Rees and Son Ano. Dom. 1826'. The cornice stones have a groove which is lead-lined to form the front gutter. The individual store rooms were rented to different merchants. The lintels of the basement windows can be seen, although the original openings were bricked up to comply with the Customs regulations for bonded stores.

Year Built	1830
Original owner	John Biddle of Stroud, miller
Architect	W. Franklin of Stroud
Principal users	John Biddle, miller
	John Weston & Company, corn merchants
	Severn & Canal Carrying Co.

The windows are larger than in most of the other warehouses and have segmented brick arches reminiscent of many cloth mills in the Stroud valley. The building originally had a hipped roof, and a change in brickwork can be seen in the present gable ends. By 1864, the foundations under the iron columns in the middle of the building had sunk by up to nine inches, and the floors and the roof had developed a corresponding sag.

SHIPTON'S WAREHOUSE

Year built	1833
Original owner	J.M. Shipton, timber merchant
Principal users	J.M. Shipton, timber merchant
	Bretherton & Pitchford, corn merchants
	S.W. Lane, corn merchant
	Severn and Canal Carrying Co.

Originally the ground floor had double headroom and an additional floor has been inserted later.

LOCK WAREHOUSE

Year built	1834
Original owner	J. & C. Sturge, corn merchants
Principal users	J. & C. Sturge, corn merchants
	Spillers and Bakers, corn merchants
	Gopsill Brown and Sons, sack hirers

The original windows and cast-iron columns were rather widely spaced, and additional columns were inserted to strengthen the building in 1877. These were supplied by the local engineers William Savory and Son and carry their name. The associated timber beams are in two overlapping pieces that are bolted together. The roof had to be replaced following damage from the fire which destroyed the adjoining warehouse in 1917.

PILLAR WAREHOUSE

Alternative name	Pillar and Lucy Warehouses
Year built	Probably 1838
Original owners	Northern half — Samuel Baker, gent. Southern half — J.M. Shipton, timber merchant
Architect	Probably S.W. Daukes of Gloucester
Principal users	J. & C. Sturge, corn merchants W.C. Lucy and Co., corn merchants Foster Brothers, oil and cake mills

The upper floors project forward and are supported by pillars on the quay wall. The windows are larger than in most of the other warehouses. The floors are supported on massive beams that are over sixty feet long.

STURGE'S WAREHOUSE

Alternative names	'G' Stores or Reynold's Double Warehouse
Year built	1840
Original owners	J. & C. Sturge, corn merchants
Architect	S.W. Daukes of Gloucester
Builder	William Rees
Principal users	J. & C. Sturge, corn merchants S.W. Lane, corn merchant G.T. Beard, warehouseman J. Reynolds & Co., millers

VINING'S WAREHOUSE

Alternative name	Reynold's Warehouse
Year built	1840
Original owner	C.J. Vining, corn merchant
Architect	T.S. Hack
Builder	William Wingate
Principal users	Vining and Rea, corn merchants J. Edwards & Co., corn merchants J. Reynolds and Co., millers

Originally the ground floor had double headroom and an additional floor has been inserted later.

166

The Lock Warehouse showing the limited number of original (small) windows.

PHILLPOTT'S WAREHOUSE

Year built	1846
Original Owner	A.H. Phillpotts, corn merchant
Architect	John Jacques of Gloucester
Builder	William Wingate
Principal users	Phillpotts & Co., corn merchants

KIMBERLEY'S WAREHOUSE

Year built	1846
Original owner	Humphrey Brown, later M.P. for Tewkesbury
Architect	Probably John Jacques of Gloucester
Principal users	J.P. Kimberley, corn merchant
	G.T. Beard, warehouseman

HERBERT'S WAREHOUSE

Alternative name	Robinson's Warehouse
Year built	1846
Original owner	Samuel Herbert, solicitor
Architect	Probably John Jacques of Gloucester
Principal users	J. & C. Sturge, corn merchants
	T. Robinson & Co., corn merchants

VICTORIA WAREHOUSE

Year built	1849
Original owner	William Partridge, merchant
Architect	Probably John Jacques of Gloucester
Builder	William Jones
Principal users	Wait James and Co., corn merchants
	S.W. Lane, corn merchant
	Turner Nott and Co., corn merchants
	Priday Metford and Co., millers

CITY FLOUR MILLS AND WAREHOUSE

Year built	1850 and probably 1854
Original owners	J. & J. Hadley, millers
Principal users	J. & J. Hadley
	Reynolds & Allen
	Priday Metford and Co.

The windows of both buildings are larger than in most of the other warehouses. The warehouse was badly damaged by a fire on 4 January 1888, and the subsequent rebuilding incorporated silos for storing the wheat in bulk. An extension containing further silos was built *circa* 1898, and a large concrete silo was added in 1964.

ALBERT WAREHOUSE

Year built	1851
Original owner	William Partridge, merchant
Architect	Probably John Jacques of Gloucester
Builder	Joseph Moss
Principal users	W.C. Lucy and Co., corn merchants
	J. Reynolds and Co., millers

The warehouse was converted to a flour mill by James Reynolds in 1869, and a boiler house etc. was built to the south. The mill closed in 1977 and the ancilliary buildings were demolished.

BRITANNIA WAREHOUSE

Year built	1861
Original owner	William Partridge, merchant
Architect	Probably John Jacques of Gloucester
Principal users	H. Adams and Co., corn merchants
	G.T. Beard, warehouseman

FOSTER BROTHERS OIL AND CAKE MILL

Year built	1862
Original owners	T.N. and R.G. Foster
Architect	George Hunt of Evesham
Builder	Eassie and Co.
Principal users	Foster Brothers
	British Oil and Cake Mills
	West Midland Farmers

The original building was considerably extended in the early 1890s. A fine two-storey wooden structure on the front, supported by pillars, had to be replaced after it was badly damaged by the collapse of the quay wall in 1892.

GREAT WESTERN WAREHOUSE

Year built	1863
Original owner	William Partridge, merchant
Principal users	W.C. Lucy and Co., corn merchants
	Fox Clinch and Co., corn merchants
	Great Western Railway Co., sugar stores
	Bristol Steam Navigation Co., sugar stores

The name Great Western was adopted long before the warehouse had any direct association with the Great Western Railway Company, and presumably refers to its size and location. The building was seriously damaged by a fire on 3 January 1945, and only the ground floor of the original building survives with a modern roof.

ALEXANDRA WAREHOUSE

Year built	1870
Original owners	J.E. and S.H. Fox, corn merchants
Builder	J. Moss
Principal users	J.E. and S.H. Fox, corn merchants
	Bristol Steam Navigation Co., sugar stores

The building was badly damaged by a fire on 21 August 1875, and it was rebuilt with a parapet instead of the original eaves where the fire started.

The Alexandra Warehouse with a parapet in place of eaves.

LLANTHONY WAREHOUSE

Year built	1873
Original owners	Wait James and Co., corn merchants
Architect	Capel N. Tripp of Gloucester
Principal users	Wait James and Co., corn merchants
	Western Trading Co., builders merchants

FOX'S MALT-HOUSE

Year built	1888
Original owner	S.H. Fox, corn merchant
Architect	J.P. Moore of Gloucester
Principal users	Fox Clinch and Co., maltsters

DOWNING'S MALT-HOUSE EXTENSION

Year built	1899
Original owners	G. & W.E. Downing, maltsters
Architect	Walter B. Wood of Gloucester
Principal users	G. & W.E. Downing, maltsters

The building by the canal was constructed as an extension to the earlier malt-house in Merchants Road. The upper floors project forward over the quay and are supported by pillars.

Chronology of Dock Developments

First Act of Parliament	1793
Work started on the Main Basin and the canal	1794
Main Basin completed	1799
River lock opened to bring the Basin into use	1812
Small graving dock constructed	1818
Barge Basin constructed	1824–5
Canal opened to Sharpness	1827
Bakers Quay constructed	1836–40
Small graving dock enlarged	1837
High Orchard Dock constructed	1840
Hempsted graving dock constructed	1846
Berry Close Dock constructed	1847
Victoria Dock constructed	1847–9
Britannia Quay constructed	1847–8
Llanthony Quay constructed	1852
Large graving dock constructed	1852–3
New quay constructed for timber yards	1854–5
Berry Close Quay constructed	1861
Sharpness New Dock constructed	1871–4
Extension of quay for timber yards	1880–2
Monk Meadow Dock constructed	1891–2
Monk Meadow timber pond constructed	1896
Wagon Works timber pond constructed	1898–9
Monk Meadow Quay constructed	1965

Annual Tonnage Statistics

	Total Tonnage (thousands of tons) Note A	Foreign Imports (thousands of tons) Note B	Foreign Exports (thousands of tons) Note B
1827	107		
1832	322		
1837	394		
1842	496		
1847	655	75	
1852	635	140	
1857	418	110	
1862	458	170	
1867	485	220	25
1872	624	325	35
1877	681	395	60
1882	643	335	80
1887	614	320	60
1892	694	365	60
1897		435	70

Note A. From Supplement to the Fifty-Sixth Annual Report of the Gloucester Chamber of Commerce (1897) G.C.L. N 15.6

Note B. Indicative values derived from various sources and including any goods which passed through Sharpness but not through Gloucester.

Principal Sources

The principal documents consulted in the preparation of this book were the minute books and other records of the Gloucester and Berkeley Canal Company and of the Sharpness New Docks and Gloucester and Birmingham Navigation Company. These are held in the Public Record Office at Kew under the references RAIL829 and RAIL864 respectively, and the specific volumes used for each chapter are noted below together with the other main sources referred to in the text.

P.R.O.	Public Record Office
G.R.O.	Gloucestershire Record Office
G.C.L.	Gloucester City Library (Gloucestershire Collection)
GJ	Gloucester Journal
GC	Gloucestershire Chronicle

Introduction

1. GJ Oct. 1784 – Jan. 1785
2. G.C.L. JR 14.6 Pamphlet announcing the proposed Gloucester and Berkeley Canal (1793)
3. GJ 12 Nov. 1792
4. 33 Geo III c97
5. Richardson A.E. Robert Mylne (1955)

Chapter One: Great Excavations

This chapter is based on P.R.O. RAIL829/1, 3, 4, and 17

1. GJ 6 Feb. 1795
2. GJ 11 Apr. 1796
3. GJ 2 Oct. 1797
4. Upton J. Observations on the Gloucester and Berkeley Canal (1815) G.C.L. JF 14.152
5. Bick D.E. The Gloucester and Cheltenham Railway (Oakwood Press 1968)
6. GJ 10 June 1811
7. GJ 29 June 1812
8. Upton J. op. cit.
9. P.R.O. BT107/153 Register of Shipping and GJ 31 Oct. 1814

Chapter Two: Completing the Canal

This chapter is based on P.R.O. RAIL829/1,2,5 and 17.

1. Upton J. *Observations on the Gloucester and Berkeley Canal* (1815) G.C.L. JF 14.152
2. G.C.L. J 14.1 Special Assembly of Proprietors (1816)
3. *GJ* 6 Mar. 1820
4. Crawford G.N. The Gloucester and Berkeley Canal Manuscripts in the Telford Collection *Glos. Soc. Ind. Arch. Journal* (1981)
5. Crawford G.N. *op. cit.*
6. *Gloucestershire Notes and Queries* Vol 4 p 317
7. G.R.O. D2460 Roll 1-F
8. *GJ* 28 Apr. 1827
9. *GJ* 18 Aug. 1827
10. Powell J.J. *Gloucestershire Extracts* 1873-8 p260
11. Counsel G.W. *History of Gloucester* (1829)
12. Shipping movements were listed each week in the *Gloucester Journal*

Chapter Three: Providing Additional Facilities

This chapter is based on P.R.O. RAIL829/2, 6 and 7

1. *GJ* 29 Dec. 1827
2. G.R.O. CA 40 Evidence given by W.B. Clegram
3. G.R.O. D2080/470 Inventory of the property of Messrs. Brown, Bradley and Harris (1834)
4. G.R.O. CA 40 *op. cit.*
5. Telford T. *Life of Thomas Telford* (1838)
6. *GJ* 1 Sep. 1832
7. *GC* 20 Aug. 1836
8. Bick D.E. *The Gloucester and Cheltenham Railway* (Oakwood Press 1968)
9. G.R.O. GDR/TI/99 Tithe map for Hempsted and South Hamlets (1840)
10. G.R.O. GDR/TI/99 *op. cit.*
11. *GJ* 22 Aug. 1840

Chapter Four: The Early Eighteen Forties

This chapter is based on P.R.O. RAIL829/8 and 9

1. Croxford J.H. The Foreign Timber Trade in Gloucester in *Supplement to the Fifty-sixth Annual Report of the Gloucester Chamber of Commerce* 1897 G.C.L. N 15.6
2. *GJ* 15 Jan. and 24 June 1842
3. *GJ* 10 Apr. 1841 and 4 Nov. 1843
4. P.R.O. RAIL37/5 Birmingham and Gloucester Railway Company
5. P.R.O. CUST42/65 H.M. Customs Surveyor of Buildings
6. *GJ* 17 Sep. 1842
7. G.C.L. SX 1.4 Memories of Samuel Aitken
8. *GJ* 15 Aug. 1840
9. *GJ* 11 Apr. 1846
10. *GJ* 29 Feb. 1840 and 5 Feb. 1842
11. Inscription on the bell and *Lloyds Register of Shipping* 1822
12. *GJ* 15 May 1841 and 24 Jan. 1846

Chapter Five: A Major Expansion

This Chapter is based on P.R.O. RAIL829/9 and 10

1. *GJ* 25 Oct. and 1 Nov. 1845 and 3 Jan. 1846
2. *GJ* 3 Oct. 1846
3. *GJ* 20 May 1848
4. Contracts for the Southgate Street Dock in the possession of British Waterways Board, Gloucester
5. *GJ* 3 June 1848
6. G.R.O. D2460 Drawing 9-B-7 and 1851 Board of Health Map PH 1086
7. *GJ* 4 Apr. 1846
8. Whalley Rev. W.H. *The Mariners Chapel* (1909) and G.C.L. N 5.7.
9. *GJ* and *GC* 21 Apr. 1849
10. *GJ* 17 Apr. 1852
11. *GJ* 12 June and 14 Aug. 1852

Chapter Six: Fluctuating Fortunes

This chapter is based on P.R.O. RAIL829/10 and 11

1. Hollins Rev. J. *Pastoral Recollections* (1857)
2. *GJ* 20 Aug. 1853
3. *GJ* 6 Aug. 1853
4. GC 11 Nov. 1854
5. Hollins Rev. J. *op.cit.*
6. *GJ* 24 Feb. 1855
7. *GJ* Oct. 1854 to Mar. 1855
8. *GJ* 21 June 1856 and 9 May 1857

Chapter Seven: 'Nowhere is Any Inertness Visible'

This chapter is based on P.R.O. RAIL829/12 and 13

1. *GJ* 24 Nov. 1860
2. *GJ* 1 Dec. 1860
3. *GJ* 3 Nov. 1860
4. G.C.L. SA 4.7 Obituary of William Clegram (1863) and *GJ* 19 Apr. 1862
5. *GJ* 13 July 1861
6. *GJ* 19 Oct. 1861
7. *GJ* 14 Nov. 1863
8. *GJ* 7 Feb., 18 Apr. and 1 Aug. 1863
9. *GJ* 9 May 1863
10. G.R.O. GBR B4/5/1 and B4/5/3 Finance and Waterworks Committee of the local Board of Health
11. *GJ* 5 Dec. 1863
12. *GJ* 13 Apr. 1867

Chapter Eight: The Challenge of Competition

This chapter is based on P.R.O. RAIL864/1 to 3 and 46

1. *GJ* 10 July 1869
2. *GJ* 4 Feb. 1871
3. *GJ* 28 Nov. 1874
4. *GJ* 28 Aug. 1875
5. Whalley Rev. W.H. *The Mariners Chapel* (1909)
6. *GJ* 9 Apr. 1887
7. *GJ* 11 July 1885
8. G.C.L. J 14.15 Extracts from Annual Reports of the Chamber of Commerce 1886–9
9. GC 8 June 1889
10. *GJ* and GC 7 Jan. 1888
11. *GJ* and GC Nov. 1889 to Sep. 1890

Chapter Nine: Around the Turn of the Century

This chapter is based on P.R.O. RAIL 864/4,5, 46 and 52

1. *GJ* 30 Apr. 1892
2. *GJ* 11 and 18 June 1898
3. *GJ* and GC 26 Aug. 1899
4. Bateson Charles *The Convict Ships 1787–1868* (1969) and *GJ* 10 Mar. 1906
5. G.R.O. D2195 Memories of Mr. D.V. Webb
6. G.R.O. GBR L22/11/7 and 8 South Ward Rate Books for 1900
7. *GJ* 1 Oct. 1904 and 14 July 1906
8. *GJ* 9 Mar. and 22 June 1907

Chapter Ten: Cargo Handling

This chapter is based on the recollections of eye-witnesses

1. *GJ* 6 Mar. 1909

Chapter Eleven: A Changing World

This chapter is based on the recollections of eye-witnesses.

1. *GJ* 15 Dec. 1917
2. *GJ* 30 Nov. 1918
3. *GJ* 15 May–5 June 1926
4. Citizen 3 and 5 Jan. 1945

Appendix One: The Warehouses

G.R.O. D2460 Original drawings of the warehouses
P.R.O. RAIL829 and RAIL864 minute books
G.R.O. Rate books for St. Owens, South Hamlets and South Ward
G.C.L. Box 8.58 Goad's Insurance Plans of Gloucester (1891)

Further Reading

Bick D.E.	The Gloucester and Cheltenham Railway (Oakwood Press 1968)
Faulkner A.H.	Tankers Knottingly (Robert Wilson 1976)
Faulkner A.H.	Severn and Canal and Cadburys (Robert Wilson 1981)
Field R.D.	The Grand Scheme (1977) G.C.L. J 14.121
Hadfield C.	The Canals of South and South East England (David and Charles 1969)
Mortimer J.	The Volunteer of Gloucester 1869-70 in Gloucestershire Community Council Local History Bulletin Spring 1983
Rowles W.	Sharpness – The Country Dock (Bailey Litho 1980)
Weaver C.P. and C.R.	The Gloucester and Sharpness Canal (1967)

Index

The numbers in italics indicate illustrations.

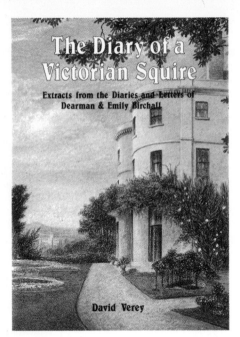

The Diary of a Victorian Squire

Dearman Birchall,
Edited by David Verey

This is the story of a Quaker cloth merchant from Leeds, who bought a country house in Gloucestershire and became integrated with the Victorian squirarchy. Magistrate, Alderman, and in due course High Sheriff, Dearman Birchall pursued the fashionable life, the season in London and the winter abroad. His brilliant wife's letters home on their six-month wedding tour, and later from Moscow and Gibraltar are outstanding features of the book. We get fascinating glimpses of their interior decorator Aldam Heaton, the aesthetic movement, and their acquaintances Matthew Arnold and Oscar Wilde, as well as the servant problem, the pleasure of tricycling, and their country neighbours Thomas Gambier Parry, St. John Ackers, Bishop Ellicott and the rest.

Squire Birchall became a pillar of the established church; at the same time he was a great china-collector, art connoisseur, theatre-goer and ladies' man. This book provides a delightful insight into everyday upper class family life in Victorian times and is edited by the squire's grandson David Verey who has written explanatory notes throughout.

224pp 219mm x 157mm
Illustrated
ISBN 0 86299 055 6 (cloth) £8.95
ISBN 0 86299 048 3 (paper) £5.95

The above prices are current (June 1984) and are subject to alteration.

The Canal Boatmen 1760-1914

Harry Hanson

Harry Hanson here discusses the economic and social condition of the Canal people against the background of the developing canal system and its eventual decline under the impact of the railway age. He offers evidence challenging existing beliefs on the origins of the boatmen and the importance of the 'Number One', and reveals when the 'family' boat first became widespread. The life style of the boatmen is studied from contemporary descriptions and he shows how a distinctive waterway sub-culture developed through the nineteenth century. An attempt is made to establish how much truth there was in the allegations that drunkenness, violence and immorality flourished on board the narrow boats.

The book will be of interest to transport and social historians of the nineteenth century as well as to the general reader. A number of rare photographs are included, together with extensive statistical appendices.

Sovereign
256pp 219mm x 157mm
Illustrated
ISBN 0 86299 067 X (paper) £5.95

The above price is current (June 1984) and is subject to alteration.

Stroudwater & Thames and Severn Canals Towpath Guide

Michael Handford and David Viner

The Cotswolds are widely recognised as one of the most attractive and visited areas in southern England. The main canals in the region, the Stroudwater and the Thames and Severn Canals, cross the Cotswolds from west to east and penetrate some of the most beautiful countryside in England, including picturesque areas such as the Golden Valley at Chalford.

Built between 1775 and 1789, the two canals are thirty-seven miles long with fifty-seven locks. Their more interesting features include the impressive group of locks climbing steeply from Stroud, and the famous Sapperton Tunnel, which at about two and a quarter miles long is the third longest canal tunnel ever built in England.

This walking guide takes its readers along the towpath pointing out the main features with the help of photographs, line illustrations and maps.

224pp 219mm x 157mm
Illustrated
ISBN 0 904387 61 5 (paper) £4.95

The above price is current (June 1984) and is subject to alteration.

THE Thames & Severn Canal

Humphrey Household

The Thames and Severn Canal

Humphrey Household

2nd Edition, revised with additional photographs.

The Thames & Severn Canal stretched from Inglesham, near Lechlade on the Upper Thames, to Stroud where it joined the Stroudwater Navigation leading to the Severn at Framilode.

It was the fulfilment of one of the earliest, perhaps the earliest, of all proposals to link British river navigations by an artificial waterway crossing the water-shed and when built in 1783-1789 it had the longest and largest tunnel in the world.

Its archives in the Gloucestershire Record Office are exceptional, probably unique, in the wealth of material bearing on the engineering and construction, the carrying trade and provision of boats, the management and operation of a canal in the late eighteenth and nineteenth centuries and in the long struggle to maintain the concern in the face of decreasing trade and falling revenue.

Some twenty years of research led to the production of the first edition of this history in 1969 and the description of that as 'a thorough, well-written and unusually wide-ranging study of a water-way' is evidently justified by the demand for a second edition, enlarged and with far more illustrations, in this year of the bicentenary of the Act incorporating the company.

258pp 219mm x 157mm
Illustrated
ISBN 0 86299 056 4 (cloth) £9.95

The above price is current (June 1984) and is subject to alteration.

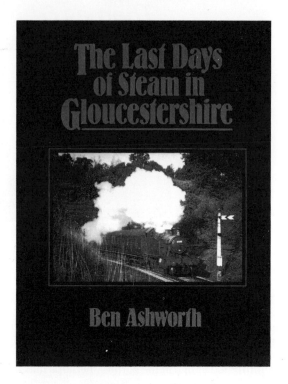

The Last Days of Steam in Gloucestershire

Ben Ashworth

This is a photographic record of the railways of Gloucestershire during a revolutionary period, between 1959 and 1966. In 1959, steam power was still pre-eminent, branch lines were still operating and stations and halts were many. By 1966, diesel power had usurped the steam locomotive, most branch lines were closed and only a small number of stations remained in use. Over 200 superbly atmospheric photographs capture this era of major change to provide a permanent record of interest to the railway buff, social historian, indeed anyone who loves steam trains.

144pp 246mm x 186mm
Illustrated
ISBN 0 86299 057 2 (cloth) £7.95

The above price is current (June 1984) and is subject to alteration.